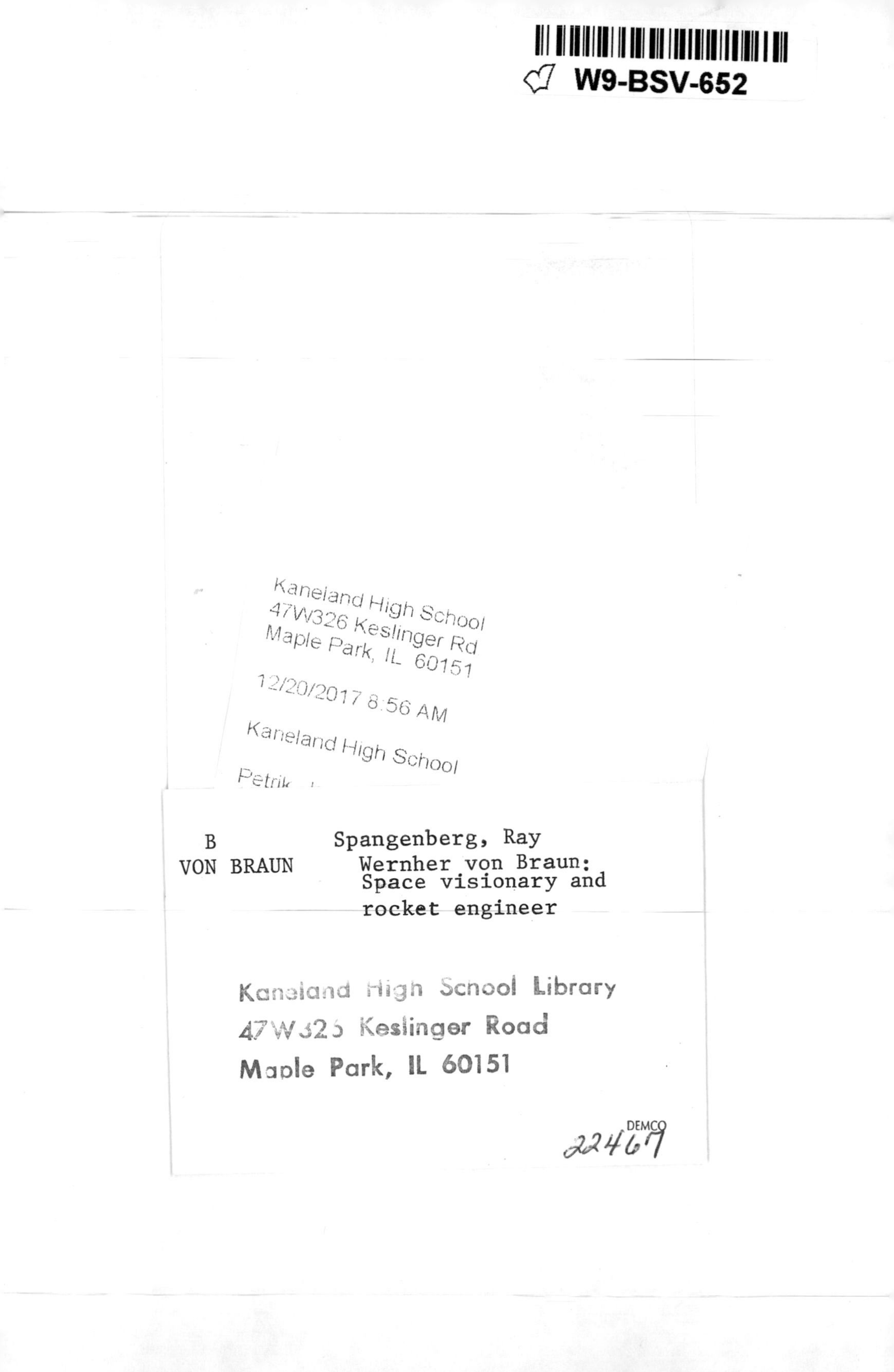

Makers of Modern Science

WERNHER VON BRAUN
Space Visionary and Rocket Engineer

Ray Spangenburg and Diane K. Moser

AN INFOBASE HOLDINGS COMPANY

To Frances Dorothy Spangenburg,
who, by her style and dignity,
has always set an example
for her son and daughter-in-law

WERNHER VON BRAUN: Space Visionary and Rocket Engineer

Facts On File, Inc.
460 Park Avenue South
New York NY 10016

Library of Congress Cataloging-in-Publication Data

Spangenburg, Ray, 1939-
Wernher von Braun: space visionary and rocket engineer/Ray Spangenburg and Diane K. Moser.
p. cm.—(Makers of modern science)
Includes bibliographical references and index.
ISBN 0-8160-2924-5 (acid-free paper)
1. Von Braun, Wernher, 1912-1977. 2. Rocketry—United States—Biography. 3. Rocketry—Germany—Biography. I. Moser, Diane, 1944- . II. Title. III. Series.
TL781.85.V6S65 1995
621.43'56'092—dc20 94-22520
[B]

Text design by Ron Monteleone
Jacket design by Catherine Hyman

Printed in the United States of America

RRD FOF 10 9 8 7 6 5 4 3 2 1

This book is printed on acid-free paper.

CONTENTS

ACKNOWLEDGMENTS

We wish to thank Charles Shows, Ward Kimball, and J. Clark Bullock for their willingness to share their personal memories of Wernher von Braun. We are also grateful for help supplied by Mike Wright, archivist, and Dominic Amatore at NASA's Marshall Space Flight Center in Huntsville, Alabama, and David R. Smith of Walt Disney Studios. And, especially, we appreciate the assistance provided by Doris Ray Hunter, consultant, U.S. Space and Rocket Center, Huntsville, Alabama. A special thanks as well to our editors, James Warren and Jeffrey Golick, at Facts On File.

PROLOGUE

The six-foot Mickey Mouse strolled down the hall, turned a corner, and poked his pointed nose through one of the many doorways lining the labyrinth of corridors. Hardly anyone inside the chaotic room took any notice, and Mickey, who must not have found what he was looking for, gave his tail a brief shake, withdrew his head, and ambled on his way.

Inside the noisy room, some two dozen people sat or stood around long tables, intently working and talking over drawings, sketches, and storyboards. Amid their clatter, chatter, and frenzied activity, the recorded strains of Andrés Segovia's classical guitar ebbed and flowed according to the changing noise level. A tall, thin young woman with bare feet, modish clothes, and an artist's flair wove in and out among the tables, gracefully pouring from a pot of black coffee. The mood throughout the room was a dynamic mixture of excitement and pressure.

"What a scene. I'd give a million dollars to spend one more day of that," one of the men in the room would recall years later. "It was a happy time, every minute of it was happy."

This was also a happy time for the tall, suntanned man near the center of the room, his blond head bent over one of the room's many drawing boards. As he spoke to the artist seated at the board, 42-year-old Wernher von Braun, former builder of Germany's notorious V-2 rockets and present guiding light of the American space rocket program, also was having the time of his life. To him, the bustle of the environment, the cooperative spirit in the air, and the intense focus of those around him were reminiscent of the teams he had led so successfully as a pioneer in rocketry.

The year was 1954, and by special arrangement—sandwiched in among his enormous rocket development responsibilities—von Braun was consulting at the Walt Disney Studios in Burbank,

California, helping Disney's top artists, writers, and production people put together a television show on the visionary world of space travel.

Wernher von Braun was a long way from his days in World War II Germany and a long way from his current home in Huntsville, Alabama. He took another sip of coffee and placed his cup back on the cluttered table ("We built that whole picture on coffee," remembered writer Charles Shows), then returned patiently to advising the young artist on the fundamentals of rocket propulsion.

"We were going to have to have a segment on Tomorrowland," television director Ward Kimball remembered. "The park [Disneyland] was about ready to open. We had plenty of footage to run for Fantasyland and true-life Adventureland. [Walt Disney] . . . had four kingdoms; one of them was Tomorrowland. He had nothing in the plan, of course, about what was going to happen tomorrow, so when I gave him those *Collier's* articles [on space exploration], he came in Monday morning and said 'This is the way to go, we don't want any science fiction on Tomorrowland. These guys sound like they know what they are doing—let's get them out and let them be part of it.' We started out with Willy Ley, we got him first and he recommended von Braun."

As Shows recalled, "Walt was the kind of guy who wanted things perfect. He would say we don't want to know just how long it takes to get to Mars, we want to know how many gallons of fuel it takes and how many hours down to the second and we want the rocket not to be a phony thing but one that would work."

Wernher von Braun built rockets that worked. That's why Walt Disney pulled every string he could to bring him to this room bustling with activity and excitement. Nibbling on a piece of cake, von Braun looked up to watch with detached amusement as two writers at a nearby table argued over the possibility of life on Mars.

"Everybody was wrapped up in space," Charles Shows later recalled. "Von Braun worked on it just like he was building the rocket to go to the Moon." Everyone wanted to see what the next step was—from the person serving coffee, to the animators, to the scriptwriters, to Walt Disney himself. "It was that exciting," according to Shows. "The place just reeked of excitement. Every-

Rocket scientist Wernher von Braun (right) with Walt Disney in 1954. (NASA Marshall Space Flight Center Archives)

body put their heart and soul into it. It was like a new hobby everybody suddenly had."

For Wernher von Braun, though, it wasn't a hobby or just another job to be working on, no matter how exciting. Instead it was another step toward fulfilling his dream—a dream of building rockets and flying to the planets that had obsessed him since childhood. Lately, though, it seemed the dream had given him a new role, one that he was growing comfortable with only slowly: He had become a celebrity and national popularizer of humanity's step into the space age. Von Braun spent much of his time writing newspaper and magazine articles like the ones in *Collier's* magazine that had caught Disney's imagination. Journalists interviewed him for articles that appeared almost daily on the newsstands. And his name had become a household word. Public speeches and government hearings occupied most of his free time. Soon he would be a television star as well, appearing alongside Walt Disney, Mickey Mouse, Donald Duck, Pluto, and the whole gang.

It was a far cry from his teen years in Germany in the early 1930s when, as a member of the amateur Society for Space Travel, von Braun had helped design and construct his first small rockets from salvaged scrap. Or from Peenemünde, where under Hitler's orders he had sent deadly missiles flying to devastate the heart of London. "America's Most Famous Rocket Scientist Outlines Trip to the Moon," newspaper headlines had recently announced. How much the world had changed. How much he had changed, and yet remained the same.

PROLOGUE NOTES

p. vii "What a scene . . ." Charles Shows, Disney scriptwriter, in an interview with the authors, December 1993.

p. viii "We were going to have . . ." Ward Kimball, Disney television director, in an interview with the authors, December 1993.

p. viii "Walt was the kind of guy . . ." Shows interview with the authors.

p. viii "Everybody was wrapped up in space . . ." Shows interview with the authors.

1

RUNAWAY ROCKETS, STARS, AND SPACE

Born on March 23, 1912 in Wirsitz, in East Prussia (now Wyrzysk, Poland), Wernher Magnus Maximilian von Braun came from a family of aristocratic landowners. The second of three children, he had an older brother, Sigismund, and a younger brother, Magnus. Wernher's father, Magnus von Braun, was a baron whose family had resided in the sleepy region of Silesia for more than three centuries. His mother, the former Emmy von Quistorp, was a woman of great culture and learning. Her family, originally from Sweden, had lived for many centuries in Mecklenburg (now in northeastern Germany) and Pomerania (now in northern Poland).

Neither Wernher von Braun's aristocratic heritage nor his rural background would play much of a role in his life, though (except, perhaps, in giving him the towering confidence he would display in his later years). A more important influence was the Germany of his childhood.

World War I began two years after Wernher was born, and he was six by the time the war had come to an end, in 1918, with Germany's defeat. Economically devastated by the war and punished for its aggressions against neighboring countries, the German nation was forced to accept responsibility for initiating the conflict. Its military was sharply curtailed and controlled, and the nation was required by the Treaty of Versailles to pay a staggering $5 billion in war debts. For the German people, with their long tradition of national pride and self-assurance, these humiliations were difficult to swallow. In the ensuing attempt to rebuild their country and its self-esteem, the German people set aside their

Wernher von Braun (center) at about age 12, with his two brothers, Sigismund (left) and Magnus. (Archives, U.S. Space and Rocket Center, Huntsville, Alabama)

long-held system of autocratic rule to try democracy for the first time—setting up a shaky constitutional government known as the Weimar Republic.

The new government found itself plagued with economic disaster and skyrocketing inflation. The von Brauns, though, were more fortunate than some, and while life was difficult, it was not unduly hard. By the early 1920s the family had moved to Berlin, where Wernher's father had become minister of agriculture and education in the new Weimar government's cabinet.

Vibrant and crowded, Berlin in the early 1920s suffered economic depression along with the rest of the nation, but the city also began to take advantage of some of the freedoms that came with Germany's new experimental democracy. Censorship was eased and long-held traditions were abandoned or questioned. The city became a thriving hub of artistic and cultural activity, of opera, theater, and filmmaking. Artists and writers flocked to Berlin, many coming in from cultures beyond the German borders, as in the case of British authors W. H. Auden and Christopher Isherwood. Musicians, dancers, and artistic "hangers-on" filled the city's streets. With the high arts also came the "lower" forms—countless bawdy "reviews," vaudeville acts, and nightclubs lit up the city's night life.

Like much of the rest of the world in the 1920s, Berlin also was exploring new ideas about the liberation of women and sex. Restraints were being cast aside; young women wore shorter skirts and danced wildly in public, while young men carried hip flasks filled with alcohol. The atmosphere was at once exhilarating and tense, a strange, dark carnival of freedom and resentment, humiliation and pride. Loud jazz music competed with militaristic jingoism, and free-thinking artists and free-wheeling "flappers" and "dandies" shared the busy streets and the frantic nights. Yet many Germans remembered the old order and yearned for the days when Germany was strong and its citizens disciplined, when the nation was powerful, respected, pure—and immune from the "decadent influences" of the rest of the world.

As a boy, Wernher von Braun (whose name is pronounced VURNER FON BROWN in German and WURNUR VAHN BRAWN in English) showed little interest in the politics or social changes of his time. He shared this indifference with many other children of the aristocratic upper-middle class who remained aloof and indifferent both to the working class and their day-to-day lives and to the "bohemians" and their liberated notions of political, sexual, and artistic freedom.

Yet there was another face of Germany in the 1920s that did capture the interest of Wernher and many of his contemporaries. The growing prospects and products of science and technology had seized the imagination of the nation, and Germany—especially Berlin—was the place to be for anyone working in physics or chemistry. There exciting breakthroughs were taking place every few weeks, and Nobel prize-winning work was the rule of the day. Albert Einstein went to Berlin to work on his theory of relativity, and other great physicists and mathematicians, including Max Planck, Wolfgang Pauli, Erwin Schrödinger, and Werner Heisenberg, gathered to study or lecture in the great German universities.

Even those who were not ready for or interested in the challenge of science were caught indirectly in its web. Many young readers pored excitedly over the novels and stories of Jules Verne and H. G. Wells—tales about adventurers who plunged to impossible depths under the sea and voyaged through space to the craters of the Moon. Around them, it seemed, technology and the fruits of science were already well on the way to providing the means for such grand adventures. The airplane was still relatively novel and adventurous, and the automobile was still as freshly exciting as a child's new toy. And with the 1920s came a new fascination: the power of rocketry and the exciting, astounding possibility that rockets could provide the means for voyages into space, to the Moon, the planets, and beyond.

As children, Wernher and his brother Sigismund tried their hands at shooting off small rockets purchased from the neighborhood fireworks stand. As Sigismund later told the story, one of them "landed in a fruit stand and ruined some perfectly good apples." The two boys were grounded after this early adventure and Sigismund abandoned his adventures in amateur rocketry, but Wernher was not discouraged so easily. Showing the kind of tenacity that would become his trademark for the rest his life, soon he launched another—which promptly landed in a bakery. He also managed to cause a king-size commotion when he attached a couple of fireworks-style rockets to a small wagon and sent it careening wildly down busy Berlin streets.

"The wagon," von Braun later recalled, "was wholly out of control and trailing a comet's tail of fire, but my rockets were performing beyond my wildest dreams." The wagon finally came to a stop when the rockets burned themselves out. Some tempers also had been ignited, though, and this time young Wernher found himself in police custody. Only his father's timely intervention saved him from a more severe reprimand, at least from official sources. Once again Wernher found himself grounded, and was strongly advised to spend more time with his schoolwork and less with his rockets.

Schoolwork didn't interest him much in those early years, though. The man who would one day explain physics and space science to the American public on Walt Disney's television program did all right in art and music, but he found most of his other classes dull and not worth the effort.

Things began to change when his mother presented him with a small telescope at the time of his confirmation (or full admission into membership in the Lutheran Church). It was an inspired idea, appealing to Wernher's profound imagination, and he soon began to spend his evenings sweeping the night sky with his telescope's small lenses.

At about the same time his father decided it was time for an educational change and sent Wernher to Ettersburg Castle, a boarding school near Weimar (southwest of Leipzig in central Germany). One of several schools in Germany known for their progressive teaching methods, Ettersburg had a unique curriculum for its time. Students spent six hours a day on traditional academic studies, but devoted the afternoons to building and farming projects that developed construction skills such as carpentry and bricklaying. The curriculum was much more to active Wernher's liking, and he still usually had an hour or two in the evening to use his telescope.

Besides reading science fiction, he also had begun to buy and read astronomy magazines. One night an advertisement caught his interest. The ad was for a book called *Die Rakete zu den Planetenräumen* (*The Rocket into Interplanetary Space*) and featured a picture of a rocket, hinting at its amazing possibilities. Wernher sent for the book immediately, and when it finally arrived weeks later, he turned the pages with tremendous excitement. Maybe it would tell him how to build a real rocket!

Excitement quickly turned to confusion and disappointment. The book was short—a slim 92 pages—and the first pages he looked at were filled with obscure mathematical formulas and physics equations! To him, it looked like total gibberish. Wernher flipped quickly to the second section. A diagram of a rocket. Good. But more physics and math—not so good. They were his worst subjects. How could he ever figure all this out?

But the third section was different. There the author—a theorist named Hermann Oberth (oh BAIRT)—outlined his ideas about how rockets could be used to escape Earth's gravitational pull and fly free through space to other planets in the Solar System. The text described the spaceship as well as supply depots in space and

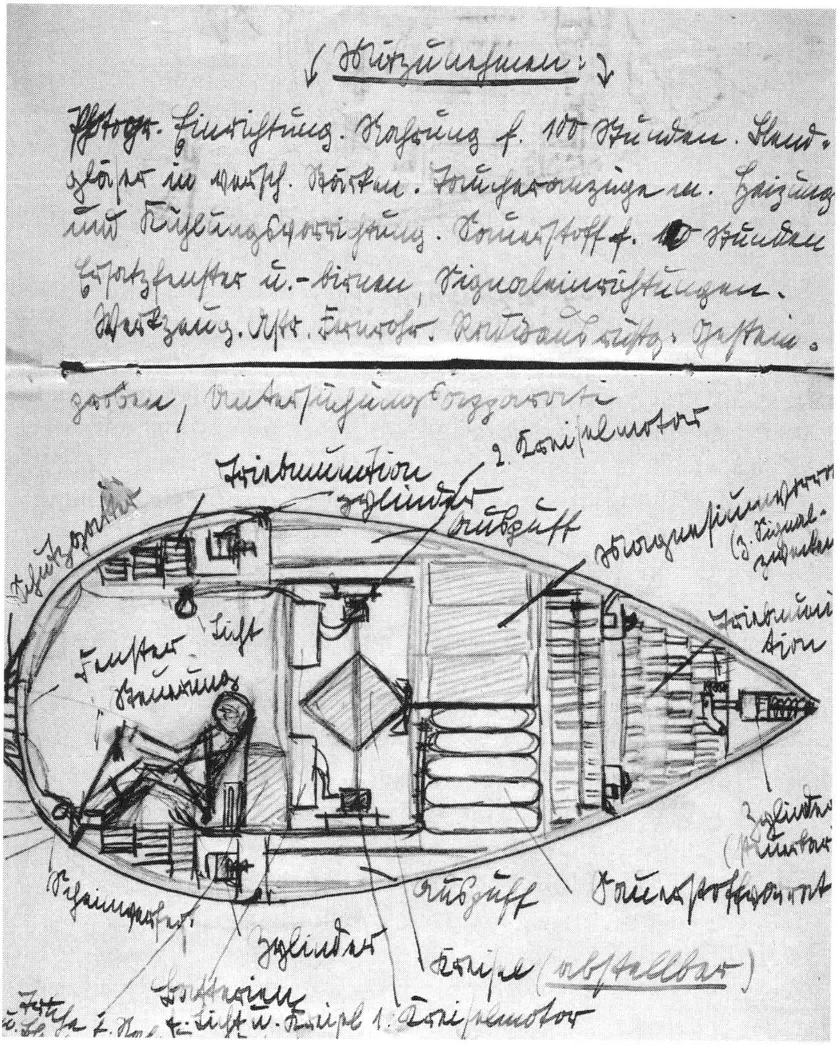

Sketch of a manned space rocket, from a page of Wernher von Braun's notebook, drawn when he was about 15 or 16. (Archives, U.S. Space and Rocket Center, Huntsville, Alabama)

the type of clothing space travelers might wear. It was like reading an outlandish fantasy, a Jules Verne or H. G. Wells science-fiction tale. Yet this book filled with mathematics and physics said that space exploration might truly be possible, some time in the not-too-distant future!

It was tremendous stuff, but a little frightening. The rockets needed to accomplish such marvels had to be much larger devices, much more complex than the simple fireworks Wernher had played with as a child. If he really hoped to build the kind of rockets that could take human passengers into space, Wernher realized that he would have to know mathematics and physics. He would also learn that his dream would require discipline and hard work. And, perhaps most important of all, he would discover what it was like to be possessed and driven by a single great idea.

To Wernher von Braun, his future path seemed inevitable. He would learn what he had to learn. He would do what he had to do. Although still in his teens, von Braun had taken his first steps into genius. He was obsessed with a single great idea, an idea from which he would never swerve and one that would control, for good and bad, all the future decisions of his life. From then on there would be no other direction for him. He would do whatever was necessary to fulfill his dream of building rockets. And someday those rockets would take human beings into space.

His dream would provoke loathing and controversy, admiration and glory. He would live it until his final days.

But he was not the first to have such a dream.

CHAPTER 1 NOTES

p. 4 "landed in a fruit stand . . ." Bergaust, *Wernher von Braun* (Washington, D.C.: National Space Institute, 1976), p. 35.

p. 4 "The wagon was wholly out of control . . ." Bergaust, p. 34.

2

"BEYOND THE LIMITS OF THE ATMOSPHERE . . ."

When Wernher von Braun first read Oberth's little book on rocket power, rockets had already been around for a long time. Their first recorded use occurred in China, where warring factions in the 13th century used "fire arrows." Not only could these arrows with simple rockets attached fly farther than ordinary arrows, but they also carried an incendiary substance that caused fires where they landed.

Within a few years, military use of rocket-powered weapons had spread throughout Asia, to Africa and Europe. In Europe and England, interest in rockets as weapons tapered off after about the 15th century. But in Asia, as late as the last half of the 1700s, any military operation always included at least one or two corps of rocket throwers.

When the rocket throwers in the army of Haidar Ali (1722–82) won battles against British troops in India in the mid-1700s, the British military suddenly gained a renewed respect for these frightening, flaring projectiles used in long-range barrages. The Indian soldiers' rockets were made of thick bamboo, about 8 to 10 feet long, attached to a tube of iron, which contained the fuse and powder. The range of these rockets may have been as great as 1.5 miles, and, while their accuracy was limited, psychologically their effect was devastating.

Around 1770, the British got hold of a few of these weapons. No one in the British army succeeded in reproducing their range, or even their less-than-perfect accuracy, until 1804, when William Congreve, a British ordnance officer, came up with some ideas for

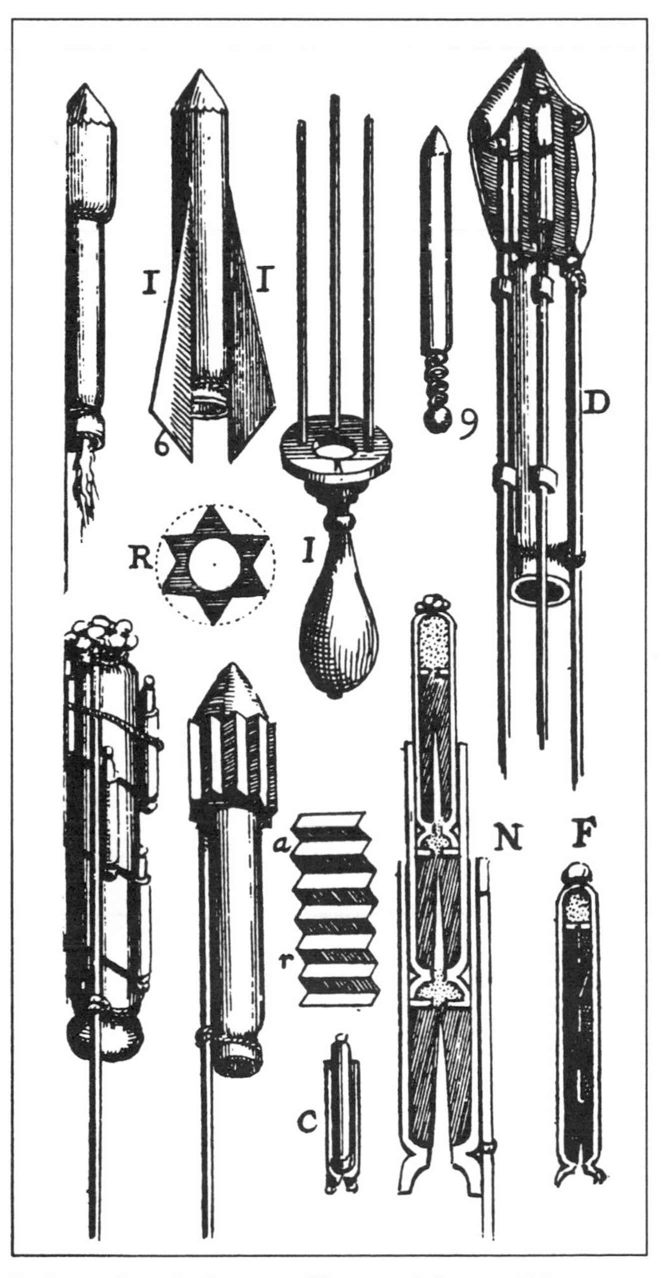

Early rocket designs, as illustrated in an 18th-century French treatise. (Library of Congress)

improving range and accuracy. Congreve also recognized an important fact: A rocket has no recoil (the push backward that occurs, for example, when a shotgun is fired). For this reason, he realized, rockets could be launched from sea, without fear of capsizing his navy's boats.

On October 8, 1806, 18 rocket boats loaded with men and officers in the Royal Marine Artillery used Congreve's improved rockets to excellent advantage against Napoleon's French invasion fleet, moored at the harbor at Boulogne, France. The following year British attacks using thousands of Congreve rockets virtually destroyed a large French fleet at anchor in the harbor of Copenhagen and, with it, much of the Danish capital city. The British also used Congreve rockets to pelt Fort McHenry from Baltimore Harbor during the War of 1812 against the United States—creating the "rockets' red glare" mentioned in the U.S. national anthem, "The Star-Spangled Banner."

Despite the 19th-century British military successes, by the 1920s rockets had taken a back seat to other weapons—the big guns used by the Germans in World War I as well as tanks and howitzers (cannon that can shoot shells both very high and very far). The big, 420-mm (16.5-inch) howitzer known as Big Bertha (or the Paris gun) could volley a 1,764-pound shell as far as 6 miles. With it the German artillery forces terrorized Paris during World War I. Rockets, as they had evolved by that time, seemed mild by comparison, and their military importance faded again. For those interested in interplanetary travel, however, rockets—which can function in the vacuum of space—offered exciting possibilities.

Rockets work on a jet-propulsion principle. When an inflated balloon is untied, jet propulsion is the force that sends it careening across the room. The air inside a sealed balloon presses against all sides equally at once. But when the neck is untied, the air pressing against the neck suddenly escapes through the hole, creating a pressure imbalance. Now the air at the top of the balloon is pressing more forcibly than the air at the neck, which is meeting almost no resistance, and the balloon is pushed in the direction of the increased pressure. (The balloon is not propelled by the air rushing out the neck, as many people think.)

People have known about jet propulsion for a long time—for example, in the 2nd century A.D., Hero of Alexandria demonstrated the power of escaping steam to move a machine he devised, and in the 17th century Isaac Newton established the principle behind this power with his Third Law of Motion: For every action there is an equal and opposite reaction.

In a rocket, burning fuels produce gases, and heat causes the gases to expand. The gases escape through an opening at one end of the rocket (the nozzle). As gas escapes from the rear, gases within the rocket press forward, propelling the whole device forward. This is what gives a rocket its traveling power.

What made rockets especially interesting to space enthusiasts like von Braun in the early 20th century was one major fact: Rocket engines don't need air to operate. Rocket engines contain within them everything they need to work—both fuel and an oxidizer (a substance that can produce the oxygen required by the fuel to ignite and burn). This means rockets can work as well in space as in the Earth's atmosphere because the oxidizer is carried either as part of the fuel or in a separate compartment.

Jet engines, by contrast, are "air-breathing" engines. Unlike rockets, they take oxygen from the air, combine it with the fuel they carry, ignite the combined mixture, and propel themselves forward from the energy produced. These engines could never be used beyond Earth's atmosphere.

So, in the early 20th century, nonmilitary uses for rockets soon began to gain ground—at least in the minds of a few individuals of mathematical talent and far-ranging vision. These theorists and engineers began to say that rockets could take us beyond the cradling atmosphere of Earth, into the cosmos. Rockets might have the power to escape Earth's gravitational pull. Rockets, in fact, might take us to the Moon.

At the time, this was an absurd-sounding proposition. But as the Russian theoretician Konstantin Eduardovitch Tsiolkovsky [tseel KOF skee] wrote in a letter in 1911, "Mankind will not remain on the earth forever, but in the pursuit of light and space will at first timidly penetrate beyond the limits of the atmosphere, and then will conquer all the space around the Sun." Both a sound theorist and a great visionary, this self-taught Russian schoolteacher was

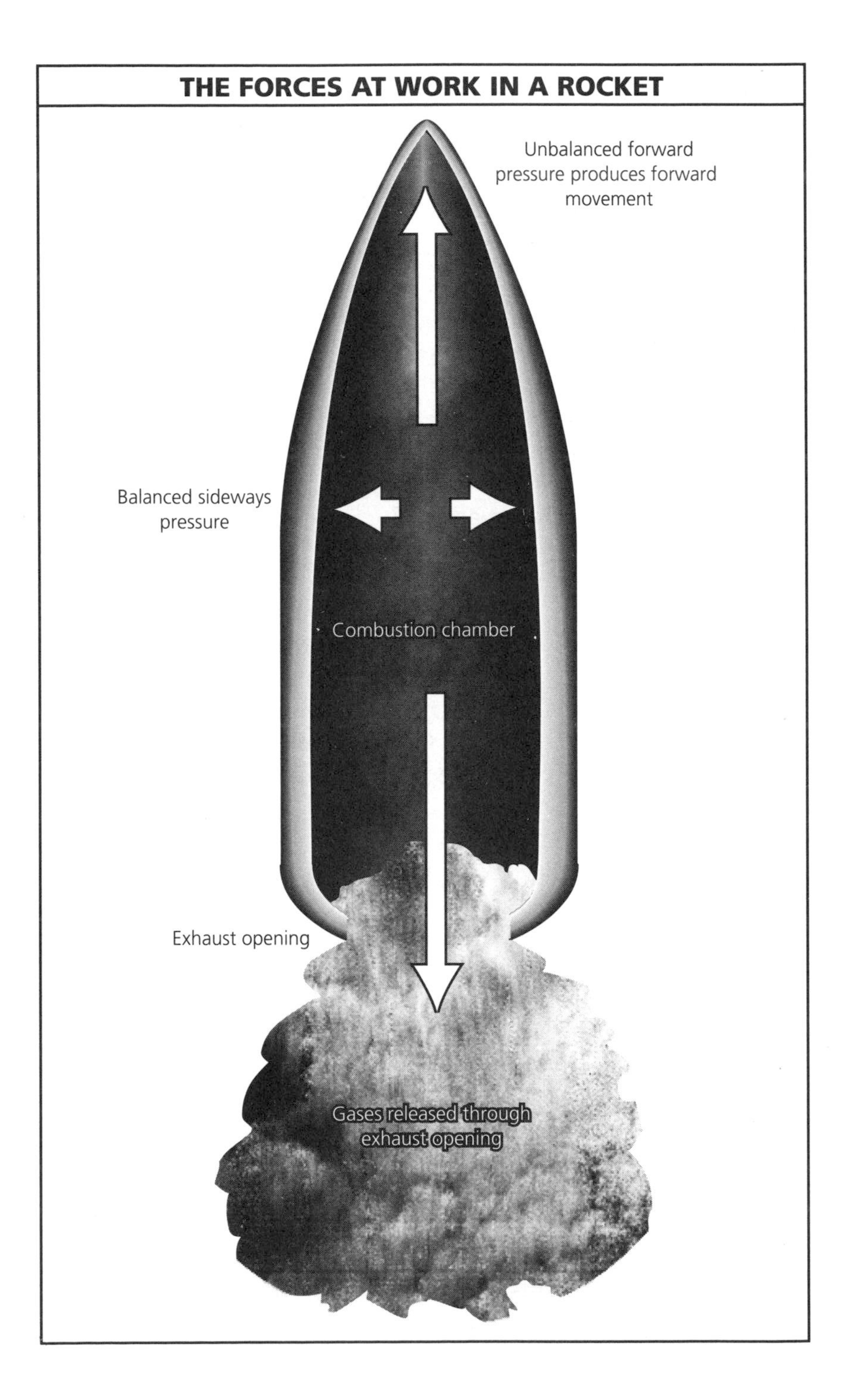
THE FORCES AT WORK IN A ROCKET
Unbalanced forward pressure produces forward movement
Balanced sideways pressure
Combustion chamber
Exhaust opening
Gases released through exhaust opening

the first to put forth the idea of rocket-powered space travel, supported by equations and detailed physical descriptions.

In his paper "Exploration of Space by Rocket Devices," written in 1889 and published in 1903, Tsiolkovsky outlined his ideas for the use of rockets in space, including a diagram of a rocket spaceship. This work became one of the most significant papers in the history of rockets and space travel.

Tsiolkovsky never actually built a rocket, but as early as 1897 he had derived his now-famous formula (still considered valid) that establishes a ratio among a rocket's speed at any point in time, the speed of the gases expelled from its engine nozzle, the mass of the rocket, and the mass of the explosives consumed. He recognized that the rate of motion in empty space is theoretically unlimited (at least, as we now know, up to the speed of light). He also saw that speed depends solely on the exhaust velocity of gas particles and the ratio of the mass of explosives (propellants) to the mass of the rocket. This idea pointed toward the possibility of attaining the enormous speeds necessary to escape Earth's atmosphere.

In the United States, meanwhile, a lone, serious-minded engineer named Robert Goddard began filing for rocket patents as early as 1914 and was testing solid-fuel rockets regularly by the 1920s. In fact, in a paper he submitted to the Smithsonian Institution in 1919—entitled blandly "A Method for Reaching Extreme Altitudes"—he suggested that it would be possible to send a rocket to the Moon!

Many people immediately jumped to the conclusion that Goddard was a fool, a crackpot with ludicrous dreams. Goddard was dismayed by the hyped-up portrayal of his work, confiding later, when asked about his Moon rocket, "All I'm trying to do is get this thing off the ground."

On March 16, 1926 Robert Goddard made an enormous breakthrough when he launched the world's first successful liquid-fuel rocket. It was an important landmark for the future of space travel, since only the liquid-fuel rocket was powerful enough to carry heavy cargoes upward beyond the edges of Earth's atmosphere. In technical terms, liquid fuels produce more pounds of thrust per pound of propellant burned per second, which means that liquid

American rocket engineer Robert Goddard (center) on September 23, 1935 with Charles Lindbergh (right, next to Goddard), who believed in Goddard's work and obtained funding for him from the Guggenheim Foundation. (Harry F. Guggenheim is next to Goddard on the left.) The group is standing in front of one of Goddard's rockets in its launch tower. (NASA Marshall Space Flight Center Archives)

fuel adds less weight to a rocket while producing more power than solid fuels.

Goddard's work, like Tsiolkovsky's, was not widely known in Europe. Yet a German-speaking schoolteacher born in Transylvania (now in Romania) also had become intrigued with both the theory behind rocket propulsion and the possibilities for space travel offered by rockets and had been in communication with Goddard.

In 1923 Hermann Oberth published the book Wernher von Braun sent for, *The Rocket into Interplanetary Space.* In it he set forth the theoretical basis for large liquid-propellant rockets, including the multistage rocket—and he compared Goddard's conclusions with his own. In later years Goddard felt that Oberth received credit for some insights that were contained in his Smithsonian paper—and without question, many conclusions were similar.

In any case, the book was an immediate popular success, directly or indirectly inspiring many whose names would soon become famous in the history of rocket development—including Rudolf Nebel, who would build some of the earliest European rockets; and Willy Ley, who would become a well-known science writer and popularizer.

CHAPTER 2 NOTES

p. 11 "Mankind will not remain . . ." K. E. Tsiolkovsky, *Selected Works,* (Moscow: Mir Publishers, 1968), p. 14.

p. 13 "All I'm trying to do . . ." Lehman, Milton. *This High Man: The Life of Robert H. Goddard,* with a preface by Charles A. Lindbergh. (New York, Farrar, Straus and Co., 1963), p. 144.

3

"HELP CREATE THE SPACESHIP!"

By the winter of 1929–30, amateur rocket builder Willy Ley had become one of the major figures of the movement to build rockets in Germany, and people interested in rocketry had begun to seek him out. Still, he was surprised when he returned home in Berlin one afternoon that winter to discover a visitor waiting for him. Ley looked back on that day in his book, *Satellites, Rockets, and Outer Space,* published in 1958: "I entered my home in Berlin and heard someone playing the piano. Since I knew that the only other family member who played the piano was absent, it had to be a visitor. It was, and I even remember what he was playing—it was the first movement of Beethoven's *Moonlight Sonata.*"

The young man sitting at the piano was Wernher von Braun, the young student who had once shown a greater talent for languages and music than for physics and mathematics. It seems he had already developed his lifelong knack for making himself at home, wherever he went. At the time, he was about 18 years old, studying at the *Charlottenburg Technische Hochschule,* a renowned technical school in Berlin.

"Physically," Ley would later note in a letter to the British Interplanetary Society "[von Braun] . . . had bright blue eyes and light blond hair and one of my female relatives compared him to the famous photograph of Lord Alfred Douglas of Oscar Wilde fame. His manners were as perfect as rigid upbringing could make them. . . ."

The reason for the memorable young man's visit became apparent quickly. He wished to join Ley and the small group of enthusiasts who had formed the amateur rocket-building society known as the VfR (*Verein für Raumschiffahrt,* or Society for Space Travel), by this time a well-established society with clearly defined goals.

Founded on July 5, 1927 in the back room of a restaurant called the Golden Scepter in Breslau (now Wroclaw, Poland), the group had a formal charter and serious objectives from the beginning. The charter set forth two principle purposes: (1) to popularize the idea of rocket flight to the moon and planets, and (2) to conduct serious experiments in the development of rocket propulsion.

It was an ambitious agenda, and the VfR was tireless in its diverse but dedicated activities—publishing posters, giving lectures, and holding public demonstrations, all of which met with enormous success, initially.

Germany, with its national ego still battered by the World War I defeat, was badly in need of some reassurance, some new confidence in its citizens' abilities. The German people fell in love with the kind of engineering perfection required by the new technologies, and they found the idea of rockets romantically adventurous. Perhaps powerful, spacebound rockets could be built. And with them, perhaps, humankind could truly conquer the stars!

By the late 1920s, rockets had become such a hot topic that Fritz Lang, an internationally acclaimed German film director, decided to make a science fiction film, called *Frau im Mond (Woman in the Moon),* about traveling to the Moon. As Walt Disney would do 30 years later with his Tomorrowland programs, Lang wanted his film to be technically correct and immediately began looking for an expert on rocketry to hire as a technical consultant.

Lang also wanted a dramatic demonstration of the possibilities of rocket travel. He hit upon the idea of building a working rocket to use in a publicity stunt, planning to shoot it 70 feet into the air above the Baltic Sea, and he turned to the man who had written the best book in German on the subject, Hermann Oberth. Eagerly Oberth agreed both to consulting on the film and to building the rocket.

Unfortunately, the plan was overly ambitious. No one in Europe had ever shot a rocket that high (although American Robert Goddard, unknown to most Europeans, had sent his liquid-fuel rocket as high as 41 feet by 1926 and 204.5 feet by the end of 1928). Also,

Lang didn't figure on the extreme difficulty of finding or manufacturing parts for rockets in 1920s Germany. And he had not, perhaps, chosen the ideal people for the job. Rudolf Nebel, who had been hired to assist Oberth, had piloted a plane during World War I and had a degree in engineering, but had never actually practiced engineering professionally. Oberth was a sound theoretician but not strong in mechanics or engineering, and the parts he designed were probably unnecessarily complicated. As a colleague once put it, "If Oberth wants to drill a hole, first he invents the drill press." The manufacturers lagged abominably behind schedule, and although Oberth and Nebel labored frantically, the publicity deadline came and went and no rocket pierced the air over the Baltic Sea. Dejected, Oberth returned home to Romania for several months.

Meanwhile, Lang's film premiered without Oberth's rocket on October 15, 1929 and was a stunning success. The movie's advertising invited filmgoers to travel on the first spaceship to the Moon and experience the agony and the ecstasy of liftoff, the thrill of weightlessness, and the excitement of the first steps on the lunar surface. The film itself promoted the drama of rocketry, with searchlights playing dramatically on the Moonship as it awaited night launch. And Lang, recognizing a need to build suspense as the moment of launch approached, first came up with the idea of counting backward toward launch time: the first countdown. The *Woman in the Moon* boosted interest in rockets and the potential of space travel more than ever. And by the end of that year, the VfR had 870 members, swelling to 1,000 the following year.

Not long afterward, Oberth, his enthusiasm renewed, returned to Berlin and von Braun had a chance to meet him through his new VfR acquaintance, Willy Ley. The young man's enthusiasm struck a note with Oberth, who was working on a department store exhibit on rocketry. Would Wernher like to help? Of course!

Von Braun spent every spare moment helping Oberth set up the exhibit. When it opened, he was on hand and stepped in with an air of authority and assurance, answering questions with the confidence of an expert. Travel to the Moon, clearly, was not far off, he declared roundly to all who would gather about and listen. The technology was almost there.

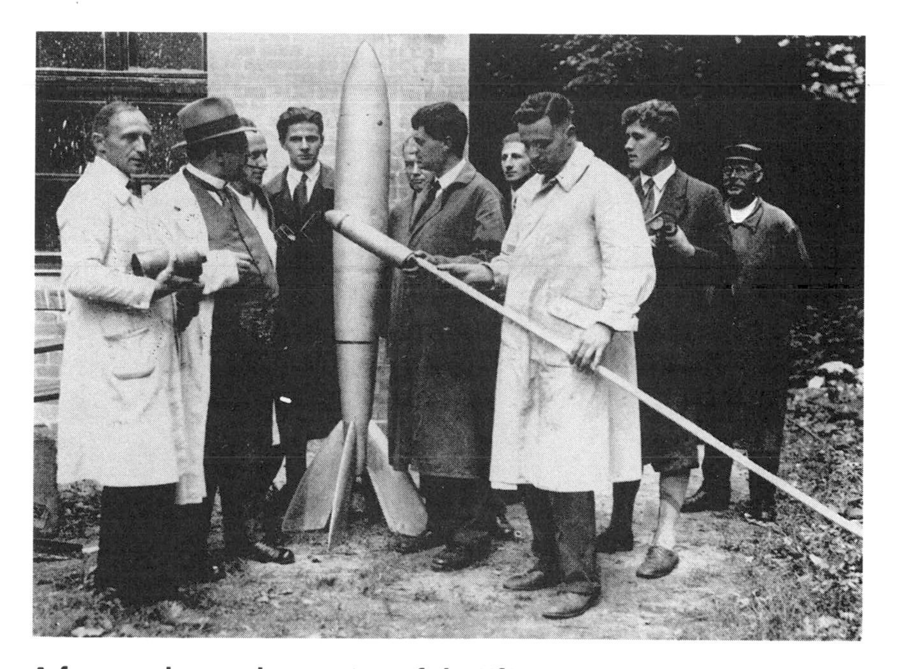

A few members and supporters of the VfR, August 5, 1930. Left to right: Rudolf Nebel, Dr. Franz Hermann Karl Ritter, Hans Bermüller, Kurt Heinisch (later in charge of a test stand at Peenemünde), an unknown enthusiast, Dr. Hermann Oberth, another unknown, Klaus Riedel (in white coat holding a Mirak I rocket), Wernher von Braun, and an unknown helper. Oberth is standing next to the partly finished model he built for Fritz Lang's film *Woman in the Moon.* (Archives, U.S. Space and Rocket Center, Huntsville, Alabama)

The *Woman in the Moon* rocket-building fiasco had produced a silver lining for the VfR and their dreams of developing working rockets. Oberth and Nebel came away with valuable experience in the realities of rocket building, and they also amassed a considerable amount of equipment, including an essential iron launching stand. Willy Ley secured the equipment for the VfR (what would Lang want with it anyway?), and Oberth and Nebel continued their work, now under the auspices of the rocket society.

The VfR's next break came when Nebel, Ley, von Braun, and several other members succeeded in obtaining permission to use an abandoned military ammunition dump in a suburb called Reinickendorf, south of Berlin. At last the group would have a place to test their rockets. They named their new test site

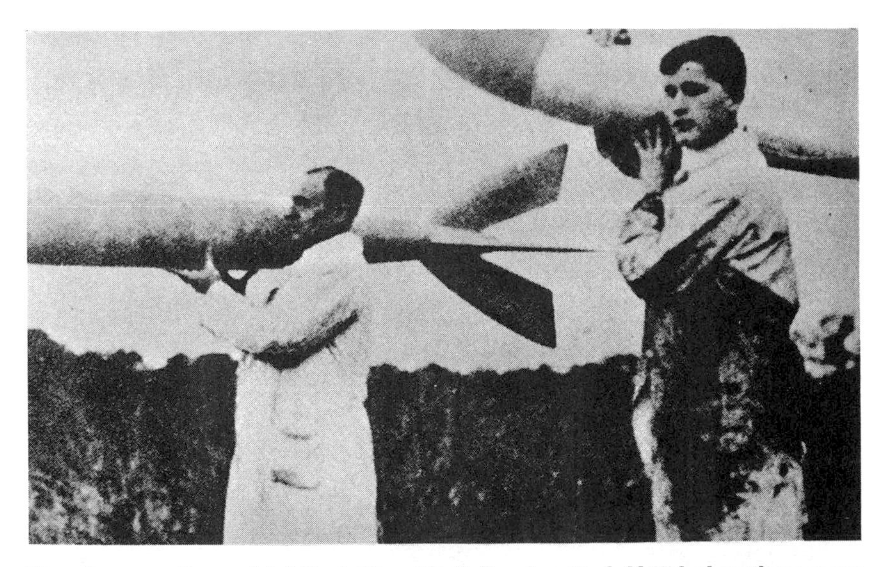

Wernher von Braun (right) at 19 or 20, following Rudolf Nebel as they carry two VfR rockets at Raketenflugplatz, about 1932. (Archives, U.S. Space and Rocket Center, Huntsville, Alabama)

Raketenflugplatz (rocket airfield) and moved in on September 27, 1930. There, in Reinickendorf, this group of ardent amateurs erected a sign reading "Raketenflugplatz Berlin" and set to work clearing the 300-acre grounds and converting the various buildings into storage areas and offices.

That first year the VfR completed 87 launchings at Raketenflugplatz, testing various rocket motors designed by its members. The group also fired an impressive 270 static firings (in which the rocket motor is fired on a test stand without allowing the rocket to take off).

While VfR members readied Raketenflugplatz, von Braun sandwiched in more technical schooling, having set off during the spring of 1931 for the *Eidgenossische Technische Hochschule* (Institute of Technology) in Zurich to continue his engineering studies. In this quaint Swiss city, nestled among the mountains at the foot of Lake Zurich, far from home and the VfR, von Braun continued to talk space exploration incessantly.

By March 1931 he had struck up a friendship with an American medical student from the University of Zurich, Constantine

Generales. During conversation one day over lunch with his new friend, von Braun embarked with ebullient enthusiasm on a discussion about future space travel and "getting to the moon."

Something about von Braun's earnestness must have saved the growing friendship from drying up that day, for "to me," Generales would later write, "this whole thing seemed very ridiculous." What turned the tide for the skeptical medical student was a letter von Braun showed him a few days later. The letter, according to Generales's account, was addressed to von Braun and signed by Albert Einstein, postmarked Berlin (although Einstein did not later recall this correspondence). As Generales recounted, "I stared at the indecipherable equations pertaining to mathematical problems and solutions in rocket design and propulsion. I was dumbfounded." If Einstein took this space travel idea seriously, Generales began to think, maybe it wasn't such a crackpot idea after all. And, while he wasn't knowledgeable about rocket engines, he did know physiology, and he began to apply his own knowledge to the issues of traveling into space.

What about human beings as space travelers? he wondered. Could the human body withstand the forces it would take to escape Earth's gravity? What would happen to muscles, organs, and bodily functions under the pressure of several gs (a g is the normal force of Earth's gravity), which would stress the body as it sped toward outer space? The answers to these questions had to be established before humans could venture into space.

"Wenn Du zum Mond gehen willst ist es viel besser zuerst mit Mausen zu versuchen!" he told von Braun pragmatically. ("If you want to get to the Moon, it is better to try with mice first!") And so began what probably were the world's first experiments in space medicine.

By July and August 1931 Generales and von Braun had built a primitive centrifuge—a bicycle wheel, bolted horizontally to a table and rigged with a belt so it could be turned with a hand crank. To its perimeter, the two students attached containers to hold the white mice they used as test subjects. With the spinning bicycle wheel they could simulate, approximately, the accelerative force that anyone aboard a spacebound rocket would experience as it launched.

"We had no idea what the tolerance of the mice would be," Generales would later recount. "In the beginning, after a few turns of the wheel, the poor mice, whose hearts you could feel pounding in the palm of your hand, were placed on the table. We observed how the little creatures would scramble slowly at first and then faster in spiraling fashion . . ."

In addition to this type of observation, Generales also dissected the mice to document what happened inside physiologically. After subjecting the mice to higher accelerations, he observed such indications of extreme stress as internal bleeding, especially in the brain, and displacement of the heart, lungs, and other organs. Using these reactions in mice as an indication of what might happen to humans, he realized that some measures would have to be taken to protect future space travelers from this much acceleration.

The two students certainly considered their efforts to be scientific research, and the unsentimental pragmatism with which they proceeded was typical of von Braun throughout his life: What do we need to know? How can we find out? Von Braun's Swiss landlady, however, did not appreciate the spatters of mouse blood on her walls and threatened him with eviction if the experiments did not stop.

So von Braun and Generales dismantled their apparatus and closed up shop. Von Braun shortly returned to Berlin and the VfR at Raketenflugplatz, where opportunity was about to knock in a very big way.

CHAPTER 3 NOTES

p. 16 "I entered my home . . ." Willy Ley, *Satellites, Rockets, and Outer Space* (New York: Signet Key Books, 1958), p. 65.

p. 16 "bright blue eyes . . ." Frederick I. Ordway and Mitchell R. Sharpe, *The Rocket Team* (New York: Thomas Crowell Publishers, 1979), p. 106.

p. 18 "If Oberth wants to drill . . ." Quoted by Ordway and Sharpe, *The Rocket Team,* p. 16.

p. 21 "to me, this whole thing . . ." Constantine Generales, "Space Medicine," in *The Coming of the Space Age,* edited by Arthur C. Clarke (New York: Meredith Press, 1967), p. 165.

p. 21 "I stared . . ." Generales, p. 165.

p. 21 *"Wenn Du zum Mond . . ."* Generales, p. 166.

p. 22 "We had no idea . . ." Generales, p. 166.

4

ROCKET ENGINEER: THE EARLY GERMAN ARMY YEARS

The excitement at Raketenflugplatz was intense, one spring morning in 1932 as members of the VfR made last-minute preparations. If things went right, their financial problems might soon be over. They had caught the attention of the Reichswehr, the army of the Weimar Republic—and a contingent from the Ordnance Department of the army was on its way to their test site to see what sorts of rockets the enthusiastic group had been working on.

Always eager to show off their rocket hardware, the members of the VfR had high expectations, as the motor car pulled at last into the grounds in Reinickendorf. One by one an austere group of solemn-faced men stepped out, dressed, not in uniform, but inconspicuously in drab civilian clothes.

With energy and excitement, the VfR's welcoming committee set about giving them the grand tour. The crisp spring air echoed with the crunch of leather shoes on gravel, in contrast to the stony silence of the army visitors. The sharpest-eyed of the strangers, Captain Walter Dornberger, took special note of every detail.

Dornberger had a particular interest in the VfR's activities. Conditions in the Treaty of Versailles had forbidden Germany from developing heavy weapons or the armed forces to use them. But, through some oversight—or simply a lack of foresight—the treaty had not mentioned rockets. And in 1929, looking for ways

Walter R. Dornberger (left), von Braun's boss at Kummersdorf, and Hermann Oberth, in a picture taken about 1942. (Archives, U.S. Space and Rocket Center, Huntsville, Alabama)

to sidestep this crippling agreement, the German army had assigned the balding, 35-year-old Dornberger to develop a liquid-fuel rocket for possible military use.

Dornberger, who held a masters of science degree in engineering at that time, had established an experimental station at the Kummersdorf army ammunition proving grounds 18 miles south of Berlin and had begun his own experiments with solid propellants. In the process, he learned firsthand some of the difficulties and dangers of rocket building. Ignoring his own safety rules, he once attempted to disassemble a solid-propellant rocket by using a steel hammer and chisel instead of safer copper tools. The sparks from the steel tools set off the rocket propellant, driving thousands of tiny black-powder particles into his uncovered face. Removing the particles had taken a year of painful visits to the military hospital, where each particle was worked out by rubbing Dornberger's face with butter and then withdrawing the particles one by one with a pair of tweezers. For members of the VfR, that morning in 1932 was just about as tense and painful.

As Dornberger would later relate in his book *V-2* and in interviews, it was obvious that the amateurs had done some impressive experimental work with rockets—but the group had no discipline, no accepted scientific procedures. They had kept no detailed records of their tests, had plotted no curves showing the degrees and range of thrust they had achieved. They had not plotted a single rocket motor's performance on a graph. In short, while the hopeful young rocket men were enthusiastic and talented enough, they also were undisciplined, unscientific, and, certainly, unmilitary in their approach.

The army was less than pleased.

As von Braun would recall in an interview shortly before his death in 1977, the army officers who visited did not like "what they called our 'circus-type' approach. We would invite people for an entrance fee to witness our launches. . . . 'What we really want,' von Braun recalled their saying, 'is data, meaningful scientific material.'"

But despite their hit-or-miss approach and their obvious lack of accepted procedure, the VfR was building liquid-fuel rockets. So, based on the low-profile, though tension-filled, official visit to Raketenflugplatz, the army awarded a 1,000-mark contract to the VfR to build a "single-stick" rocket, which the VfR called Mirak II.

The VfR contended with two major problems in their rocket development—overheating and lack of stability. By 1931, they had added aluminum walls cooled by water to the combustion chamber, which greatly cut down on explosions caused by overheating during combustion. With Mirak II, the group hoped that adding a single stabilizing stick to the tail would provide the even vertical flight they needed. It was a simple modification, and Mirak II was soon ready for demonstration. Von Braun later described its launch, in July 1932, at the army's proving ground at Kummersdorf:

> Early one beautiful July morning in 1932 we loaded our two available cars and set out for Kummersdorf. . . . As the clock struck five, our leading car with a launching rack containing the silver-painted Mirak II atop and followed by its companion vehicle, bearing liquid oxygen,

> gasoline, and tools, encountered Captain Dornberger at the rendezvous in the forests south of Berlin. Dornberger guided us to an isolated spot on the artillery range where were set up a formidable array of phototheodolites, ballistic cameras, and chronographs—instruments of whose very existence we had theretofore been unaware.

By 2:00 P.M. the rocket was fueled up and ready to go. The moment of liftoff came, and the signal was given. The rocket gave a belch and soared, reaching a height of 200 feet or more. But suddenly it became apparent something was wrong. The rocket leveled out, following a nearly horizontal path, and finally crashed before the parachute could open.

Overall, the army was not charmed by the VfR. These men knew more about liquid-fuel rocketry than just about anyone else in Germany, but clearly they couldn't build a rocket you could depend on.

Still, Dornberger did not come away emptyhanded. If he could not contract successfully with the VfR to build a viable rocket, he still might make use of the expertise of their best members. His attention settled on one who stood out from all the rest. As Dornberger later wrote in *V-2:*

> I had been struck during my casual visits to Reinickendorf by the energy and shrewdness with which this tall, fair young student with the broad massive chin went to work, and by his astonishing theoretical knowledge. . . . When General [then Colonel] Becker later decided to approve our army establishment for liquid-propellant rockets, I put Wernher von Braun first on my list of proposed technical assistants.

It was the beginning of what would become a long and challenging adventure for Wernher von Braun, one that would take him into the disastrous infamy of war and the triumphant glory of space.

Von Braun was called in by Colonel Karl Becker, Dornberger's boss, to meet with him at the Army Weapons Office, and they struck up an immediate rapport. Becker offered von Braun a civilian position developing rockets for the army, and he further insisted that von Braun pursue a Ph.D. in physics at the University of Berlin, working under Erich Schumann, who also headed the research section of the army's Ordnance Department. Von Braun could use the experimental resources at Kummersdorf to complete his thesis work, which would deal with liquid-fuel rocket engines.

To von Braun it was an unbelievable break. He hurried back to the VfR with the news—the army was interested in research done under its sponsorship and direction, on its grounds. Nebel scoffed. He had no interest in succumbing to military priorities, and other key VfR members shared his aversion. But by now von Braun had concluded that rocketry had gone just about as far as it could in the hands of amateurs with limited funding. Real progress—the kind needed for reaching space—would require more discipline and much more money. So on October 1, 1932 Wernher von Braun became a civilian employee of the German army. On that day, for him, rocketry ceased to be a purely amateur enterprise and became a professional mandate for developing a military weapon.

In an article von Braun wrote 24 years later, he asserted, "It is, perhaps, apropos to mention that at that time none of us thought of the havoc which rockets would eventually wreak as weapons of war. . . . To us Hitler was still only a pompous fool with a Charlie Chaplin moustache."

Nebel, meanwhile, turned out to be wrong that the VfR could hold out, continuing its work on its own terms. By the spring of 1933 forces beyond the VfR's control began to close in on them. The Luftwaffe turned up one day to announce that the air force would be using Raketenflugplatz as a drill ground. Then the group received a utility bill for 1,600 marks—apparently because various water pipes on the property had been leaking for years and accruing charges. The VfR could not pay and the group was turned out. In any case, by 1933 the Nazis disbanded all rocket development not sponsored by the government. The VfR's exciting contribution to rocketry had come to an end.

At the age of 21, Wernher von Braun moved into the army installation at Kummersdorf West, where Dornberger supplied him with a single rocket test stand and three colleagues, Walter Riedel, an engineer from the Association for the Utilization of Industrial Gases; Arthur Rudolph, an engineer; and Heinrich Grünow, a master mechanic from the VfR's Raketenflugplatz.

Immediately the four of them set to work, on a shoestring budget. Because rocket development by the army was a violation of the Versailles Treaty in spirit, if not in actual fact, the entire operation had to be carried out in secret, and nothing about the acquisitions for the group could look suspicious, even to auditors within the army. So Dornberger was forbidden to order any office equipment or machining tools. As he later would relate:

> We acquired things "as per sample." For instance, even the keenest Budget Bureau official could not suspect that "Appliance for milling wooden dowels up to 10 millimeters in diameter, as per sample" meant a pencil sharpener, or the "Instrument for recording test data with rotating roller as per sample" meant a typewriter . . . And if there was nothing else to do, we entrenched ourselves behind the magic word "secret." There, the Budget Bureau was powerless.

Slowly they were able to build a facility that was usable.

The first rocket engine built by the Kummersdorf team was a liquid-propellant motor using ethyl alcohol and liquid oxygen. They worked fast and by December 21, 1932, it was ready for its first static firing test.

Over the centuries, people have used the word "rocket" interchangeably to mean both an object powered by rocket power and the engine that powers the object. In recent years, to avoid confusion, it has become more common to talk about the "rocket engine" in discussing the power source that propels the vehicle, leaving the word "rocket" to mean the vehicle propelled by a rocket engine. A rocket can carry a cargo as well as its fuel and engine—and that cargo may be a warhead, in the case of a weapon, or scientific equipment, or anything else that the rocket engine is powerful enough to transport. The more powerful the rocket engine, the bigger the cargo it can propel, and the farther it can take it.

This rocket engine could not have carried much of a cargo. It was only 50.8 centimeters (about 20 inches) long, consisting of a pear-shape combustion chamber made of aluminum. The team had placed it on a stand in the center of a testing room, lit by floodlights. Walter Riedel was in charge of the valves that controlled the flow of fuel and liquid oxygen (LOX) to the engine's combustion chamber. Located behind the test site's 13-foot concrete walls, which were covered with a movable roof of tar paper and wood, Riedel probably stood in the safest position. Heinrich Grünow stood nearby, with his eye on a pressure gauge. His job was to control the valve that governed the pressure in the fuel tanks. To von Braun, standing at the open end of the test bay, fell the unenviable task of igniting the engine, using the most primitive and dangerous of methods—an "igniter" consisting of a 13-foot wooden pole with a can of flaming gasoline attached at one end. Dornberger watched from a distance of little more than 30 yards, protected by the slender trunk of a pine tree.

The orders echoed in the night air: *"Feuer! Benzin! Sauerstoff!"* (Fire! Gasoline! Oxygen!). Von Braun swung the flaming can of gasoline under the rocket nozzle, while Riedel opened the fuel and oxygen valves.

An eerie white cloud seeped from the nozzle and spread to the ground, followed by a slender trickle of alcohol. Suddenly it came in contact with the flames from the gasoline can, ignited with a *swoosh* and a hiss, then a loud crash. As Dornberger described it:

> Clouds of smoke rose. A single flame darted briefly upward and vanished. Cables, boards, metal sheeting, fragments of steel and aluminum flew whistling through the air. The searchlights went out.
>
> Silence.
>
> In the suddenly darkened pit of the testing room a milky, slimy mixture of alcohol and oxygen burned spasmodically with flames of different shapes and sizes, occasionally crackling and detonating like fireworks. Steam hissed. Cables were on fire in a hundred places. Thick, black, stinging fumes of burning rubber filled the air. Von Braun and I stared at each other wide-eyed. We were uninjured.

They were lucky. The test stand was demolished, metal doors torn off hinges, steel girders and pillars twisted and bent. But no one was hurt, although shards of sharp steel were embedded high in the trees overhead. It was their first effort, and their first failure. Many more explosions, failures, and disappointments would follow in the years to come—they were working on the cutting edge of a new and complex technology, and many factors had to balance, through trial and error, before success would be theirs.

Three weeks later a new rocket design was ready to test. After igniting successfully, the whole rocket engine burned with a white-hot heat: The aluminum of the combustion chamber jacket had caught fire. It was the Kummersdorf team's first cooling problem.

Under Dornberger's guidance, the team also tested combustion chamber designs and fuel-injection systems. The fuel injector on a rocket works a lot like a fuel injector on an automobile, spewing a fine spray of fuel or LOX into the combustion chamber, where it ignites. The fineness of the sprays and the mixture of fuel to LOX had to be exactly right, and trial and error were the only teachers in this new field.

The team also tested the flow and mixture ratios of fuel and LOX as well as various types and combinations of fuel. They measured the temperature of the exhaust flame, and they sampled the exhaust to determine its makeup. Under Dornberger's meticulous guidance, von Braun, Riedel, Rudolph, Grünow, and the others who joined them had begun to do rocket engineering in earnest.

Finally the time had come to put together their first rocket for flight, an *aggregat* that is, a rocket-engine assembly. For the *Aggregat 1,* known as the A-1 for short, the objective was to build a liquid-fuel rocket that would fly reliably and hold to a prescribed flight path. It was a tall order.

The men knew they would have to deal with the issues that had kept the Mirak II from flying straight, but no one in Germany knew any more about it than they did. For some reason, the design for the A-1 included no fins. Probably the best explanation for this design decision is that the army and Dornberger—an artillery expert—thought of the rocket as a gigantic flying shell. A shell is stabilized by its spin in flight. But spinning the entire body of a rocket would send the liquid fuel spinning around the walls of the tanks, making control of its flow into the combustion chamber almost impossible. So, at Dornberger's suggestion, the team members decided to put a gyroscopic spin on part of the rocket, to be located in the nose cone. The rotating nose section of the A-1 weighed 85 pounds, without its payload (cargo). The entire rocket measured only 55 inches and was just under a foot in diameter. In two tanks it carried alcohol and liquid oxygen, with a third tank for pressurized nitrogen, which the team used to force the propellants into the combustion chamber.

In more sophisticated systems developed later, a gyroscope would often be used to provide a frame of reference (keeping track of which direction is up) for a guidance system, which in turn uses

A-2 rocket on test stand at Kummersdorf, 1935. (Archives, U.S. Space and Rocket Center, Huntsville, Alabama)

robotlike devices to move vanes and fins to adjust the flight path and steer the rocket. In this case, the spinning section of the rocket was supposed to do the whole job all by itself, simply by the force of its spin.

It didn't work. It's easy now to scoff and wonder why the scientists didn't at least add the obvious stabilizing factor that fins would provide. Arrows use feathers to fly a straight path. Why not fins along the shaft of a rocket? But these men were working on the front lines of this technology, trying what no one had ever tried before.

Von Braun and the team scrapped the A-1 and went on to a new design, *Aggregat 2*. This rocket, which was the same size as the A-1, used a new engine design that could deliver 2,205 pounds of thrust, compared to the puny 661 pounds supplied by the A-1's engine. Von Braun and Rudolph, at least, were still thinking in terms of developing rockets for space travel, even though they had been hired to develop weapons. For that, they would have to move up to bigger and bigger engines and, ultimately, bigger and bigger rockets. In the A-2, von Braun also had moved the gyroscopic portion of the rocket—still functioning on the same principle as before—from the nose cone to the midsection. Now their rocket would no longer be top-heavy.

The team built two A-2s, which they named Max and Moritz, after the Katzenjammer Kids, popular comic strip characters at the time. They took Max and Moritz to the island of Borkum in the North Sea for launch. And they flew! Both reached an altitude of about 1.2 miles (6,500 feet); von Braun, Rudolph, Riedel, and even Dornberger were jubilant.

Meanwhile, trouble for Germany and for the world at large had been brewing. The Nationalist Socialist Party (or Nazis) had begun to gain more and more ground among the discontented electorate in Germany. By 1930 the Nazis had gained 18 percent of the vote in the elections to the Reichstag (the German equivalent of Congress or Parliament), their strength second only to the Social Democrats. On January 28, 1933 the chancellor, Kurt von Schleicher, resigned, after unsuccessfully attempting to form a coalition government, and two days later the Nazi party's leader, Adolf Hitler, was appointed to take his place. Less than a month later 40,000 men were sworn in as auxiliary policemen. Five days after that, on February 27, a fire broke out in the building that housed the Reichstag in Berlin. It was all the excuse that Hitler needed. (And, in fact, some historians claim his henchmen set the fire.) He blamed the Communists and dissidents for this attack on the heart of German democracy, and the following day he was given emergency powers that virtually made him dictator of Germany.

The Nazis won nearly 45 percent of the seats in the Reichstag in the elections held a few days later, and in the following days Hitler began to gather around him one of the most deadly groups of men in world history. Step by step he cemented his position of power, with the formation of the soon-dreaded SS (Schutzstaffel, or Black Shirts, a quasi-military police arm with enormous power); the establishment of special courts for the prosecution of political enemies; his stripping of power from individual states; the national boycott of Jewish businesses and professionals; and the formation of the Gestapo, his secret police force. On October 14 Germany withdrew from the League of Nations, signaling an end to the nation's attempt to bend to the demands of the world community. Before long it became clear that Germany was headed for war.

Under the Nazis, the army paid less and less attention to provisions of the Versailles Treaty and began openly developing weapons such as big guns and cannons. But the activity at Kummersdorf also began to step up. Although no longer needed as a way around the Versailles Treaty, the ballistic missiles that von Braun and his team were developing could serve a purpose that long-range artillery could not.

Meanwhile, the army was not the only German military service that saw uses for rocketry. While the army saw rockets as sort of huge artillery shells, the Luftwaffe (the German air force) became interested in adding rocket power to its propeller-powered airplanes, and turned to the army for help. Dornberger and von Braun saw no problem with this plan—they signed a contract, took on the job, and von Braun set about creating a design.

Static testing began in 1935, and by 1936 the work was ready for testing with a human pilot aboard. The Junkers aircraft company delivered a small, wingless fuselage, called a "Junior," to Kummersdorf. The Kummersdorf team mounted the "Junior" on a huge centrifuge and installed a rocket motor with a thrust of 300 kilograms (661 pounds) on its belly. On the other end of the long arm the scientists installed a counterweight.

Von Braun, always fascinated with flying, was especially drawn to this project. He had begun his first gliding in 1931, taking lessons at the famous Wolf Hirth's Soaring School in Grunau in Silesia. By 1933 he also had taken flying lessons and had become a licensed pilot. Later, from 1936 to 1938, he would serve two hitches in the military as a pilot in the prewar Luftwaffe, when he received pilot's ratings on heavier fighter planes, such as the Stuka and the Messerschmitt 109, and for multiengine aircraft.

So when a human pilot was needed to test the Luftwaffe's "Junior," von Braun climbed into the cockpit, strapped himself in, and signaled ignition. After he whirled around at a breathless 5 gs, the spinning stopped and he climbed down. Although pale and dizzy, he was enthralled at the success. The Kummersdorf team had achieved another milestone. They had proved that rocket power could be used with human-piloted aircraft for bursts of power and speed.

In about 1935, the Luftwaffe also became interested in an extension of this idea. Often a heavily loaded propeller-driven airplane would have trouble taking off with its cargo, although it would be fine once it was airborne. The Luftwaffe asked the Kummersdorf team to design and build rocket engines to give a heavily laden plane the boost it needed on takeoff. But to build and test on this scale, von Braun saw that they would need more room. So he and Arthur Rudolph went to the German Air Ministry with a proposal. The army and the air force, they said, should establish a joint rocket development base that would have the kind of space and facilities required by the new projects. To their delight, the idea was greeted warmly.

And so Dornberger and von Braun began looking for a bigger, better place to build and test rockets—somewhere isolated and flat, with lots of wide, open space, to minimize the danger to nearby citizens. Somewhere on the coast, where test rockets could be shot over the water and parallel along the coast so that their progress could be monitored visually and reported by radio; somewhere that could be kept secret.

While visiting his parents at the family estate in Silesia in Christmas 1935, von Braun mentioned the search. Why not look on the island of Usedom, near Peenemünde, his mother suggested, where your grandfather used to hunt ducks? The idea seemed worth looking into.

CHAPTER 4 NOTES

p. 26 "what they called our 'circus-type' approach . . ." Wernher von Braun, from an interview taped in 1976, shortly before his death, and aired in "Spaceflight," a multipart documentary shown on public television in 1986.

p. 26 "Early one beautiful July . . ." von Braun, "German Rocketry," in *The Coming of the Space Age,* edited by Arthur C. Clarke (New York: Meredith Press, 1967), p. 38.

p. 27 "I had been struck . . ." Walter Dornberger, *V-2* (New York: Viking Press, 1954), p. 27.

p. 28 "It is, perhaps, apropos . . ." Von Braun, "German Rocketry," in *The Coming of the Space Age,* edited by Arthur C. Clarke, p. 39.

p. 29 "We acquired things 'as per sample.'" Dornberger, *V-2,* p. 37.

p. 30 "Clouds of smoke rose . . ." Dornberger, *V-2,* p. 26.

5

TRIUMPH AT PEENEMÜNDE: GRAZING THE FRINGES OF SPACE

Peenemünde was located on a spit of land jutting out into the Baltic Sea from the northwestern tip of Usedom Island, due north of Berlin. Peaceful and isolated, it had few inhabitants—just acres of pine woods, rolling sand dunes, and marshes populated by many varieties of water fowl. Von Braun fell in love with the place immediately and brought Dornberger around for a look. Hidden away, yet with a clear shot along the coastline, it was perfect for the needs of the Kummersdorf team.

Usedom, along with the neighboring island of Wollin, extends along the northern edge of Stettin Lagoon, separating its waters from the Baltic Sea. From the mainland, four rivers empty their waters in that area and wend their way around the sprawling islands and into the Baltic: the great Oder River to the east, the Randow, the Uecker, and the Peene, which empties toward the western end of Usedom and flows in a channel between the island and the mainland, northwest to the sea. Near its mouth lies the tiny village of Peenemünde, literally, "mouth of the Peene."

Dornberger saw that the Kummersdorf research might profit from the rivalry between the army and the newly formed air force. The Luftwaffe wanted rocket-powered take-off capability for its airplanes. The army, in turn, wanted to offer a long-range artillery capability as an alternative to air force bombing missions, which risked the loss of valuable pilots and crew members, required extensive training, and ran up the bills.

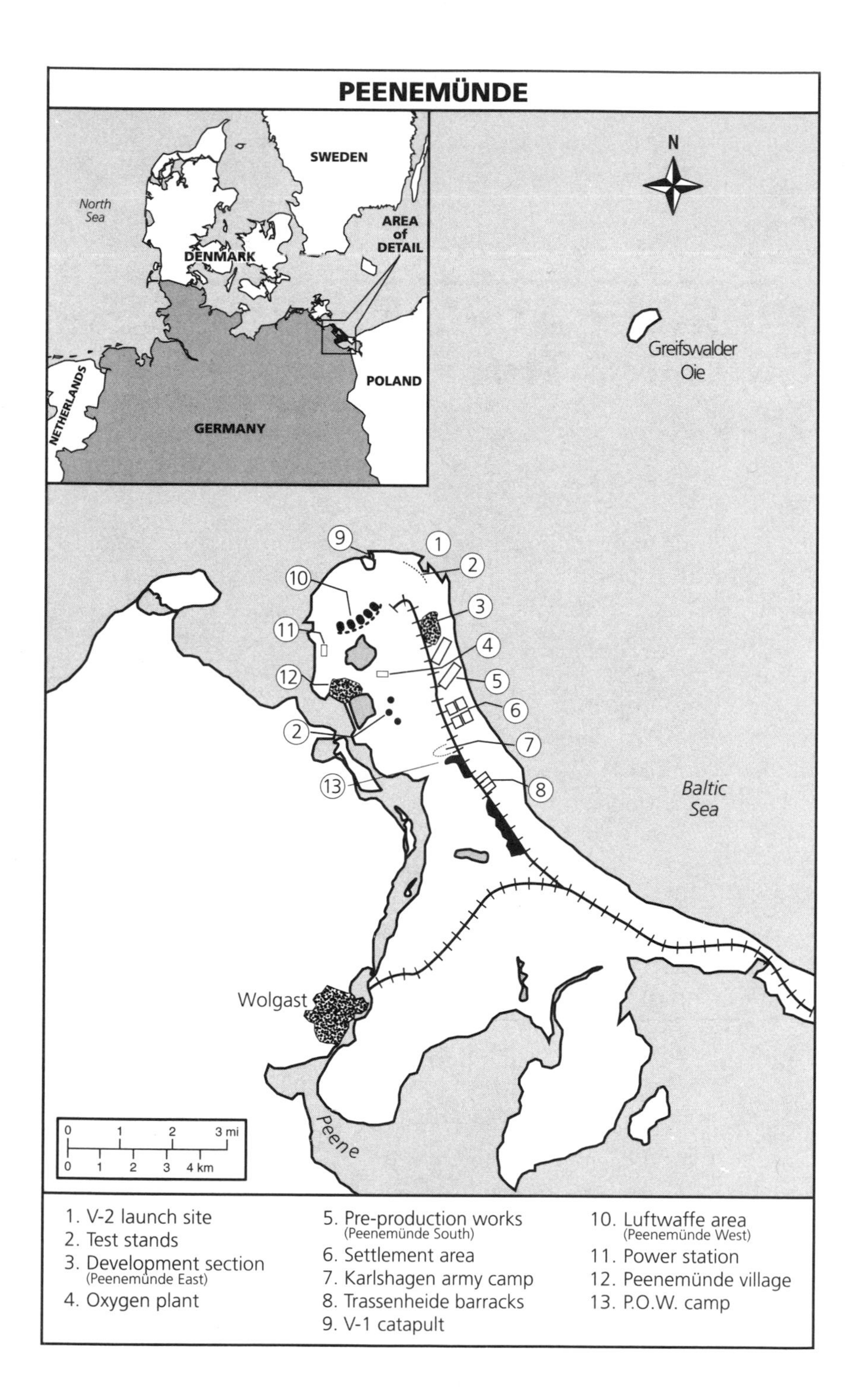
PEENEMÜNDE
N
SWEDEN
North Sea
AREA of DETAIL
DENMARK
POLAND
NETHERLANDS
GERMANY
Greifswalder Oie
Baltic Sea
Wolgast
Peene
0 1 2 3 mi
0 1 2 3 4 km
1. V-2 launch site
2. Test stands
3. Development section (Peenemünde East)
4. Oxygen plant
5. Pre-production works (Peenemünde South)
6. Settlement area
7. Karlshagen army camp
8. Trassenheide barracks
9. V-1 catapult
10. Luftwaffe area (Peenemünde West)
11. Power station
12. Peenemünde village
13. P.O.W. camp

Dornberger and von Braun played this advantage to the hilt. In March 1936 Dornberger invited Major General Wernher von Fritsch to visit Kummersdorf. Von Fritsch was commander in chief of the Reichswehr. If he was impressed, they thought, support could be theirs. When he and his staff arrived, Dornberger and von Braun went into action, with von Braun making the kind of persuasive presentation that would become his trademark. He had colored charts and graphs; he had models; he had diagrams; and he had his charismatic personality. Never known to be struck with stage fright, von Braun had an extraordinary power to gain and hold the attention of his listeners, pull them into his world and fill them with his enthusiasm. Throughout his career he mesmerized diplomats, business tycoons, presidents, and generals.

After the verbal presentation, Dornberger and von Braun took the general on a tour of the rocket test stands, where they demonstrated the Kummersdorf team's 660-pound (300-kilogram), 2,200-pound (1,000-kilogram), and 3,300-pound (1,500-kilogram) liquid- fuel rocket motors in static tests. The sheer noise and power of the demonstrations struck awe, and von Fritsch succumbed. "How much money do you want?" he asked. It was von Braun's first big success in the arena of political persuasion.

The Wehrmacht and the Luftwaffe held a joint meeting in Berlin in 1936 and, in what Dornberger later called an "attack of acute generosity," General Albert Kesselring, chief of aircraft construction, gave his approval for the purchase of the island of Usedom for use by the two services. The Luftwaffe agreed to build a rocket test site at Peenemünde, with two sections, Peenemünde-East for the army and Peenemünde-West for the air force. The Wehrmacht and the Luftwaffe would split the expenses of operating the site.

By the end of 1936 the quiet of Peenemünde was broken by the sounds of men sawing down trees, steam shovels digging out roots, and huge earth-moving machines leveling the dirt for construction. The great rocket development center had begun to take shape. By the time it was finished in 1939, it would have its own power plant, an oxygen-generating plant, test stands, bunkers, barracks, and a wind tunnel for testing the aerodynamics of rocket designs. To the south, the Pre-Production Works were housed in two big

Officers of the German army visit the facilities at Peenemünde. Von Braun is the civilian at the center, wearing a trench coat and rain hat. (Archives, U.S. Space and Rocket Center, Huntsville, Alabama)

concrete sheds, nestled among the evergreen trees and camouflaged by netting.

On the western side of the island, the Luftwaffe built its "Peenemünde West," where the development would continue on the air force's Fieseler Fi-103, an unmanned cruise missile or plane, which later became known as the V-1, or "buzz-bomb." There, also, the army's rocket team would continue building several rocket motors for assisting Luftwaffe airplane take-offs.

"Peenemünde East" would become the development site for the ballistic rockets that the army's team had begun to develop at Kummersdorf. By 1942 Dornberger would have more than 1,960 scientists, engineers, and technicians under his command as well as another 3,852 support personnel. By mid-August 1943 there would be 17,000 personnel altogether. And the military would pour more than 300 million deutschmarks (about $70 million) into the construction at Peenemünde.

There von Braun and the rocket team would complete their work on the world's first long-range ballistic missile. It would become the first weapon of its kind designed for mass production on an assembly line. It was the first truly large missile, incorporating engineering advances that would provide the prototype for big rockets developed after the war. The Peenemünde team's aim was to develop a military weapon, but the men knew this work would become the basis not only for intercontinental ballistic missiles (the ICBM) but for spacebound rockets as well. They were working on the vanguard of rocket technology, in a world all their own. And the pressure was cranking up.

Meanwhile, back at Kummersdorf, as the A-2s were being readied for testing, the rocket engine team had been working on an even more powerful engine, the 1,500-kilogram (3,300-pound) motor, which had 50 times the thrust of any rocket seen up to that time. With the financial backing of the military, the German team had moved far ahead of similar developments in the rest of the world. This new engine placed the dream of flying into space within reach.

By December 1937 the A-3 was ready for a test flight. Despite wretched weather, von Braun and members of his team headed out to the tiny island of Greifswalder Oie, north of the new rocket center being built at Usedom. There they prepared to launch three test rockets. They had no launch tower, just a bare concrete surface, and they had no range control facilities for tracking the missiles once launched. So they spent days waiting for everyone to get into position—radio operators to track the launched missile, recovery aircraft, and divers. Finally on December 4, 1937 all was ready.

The first A-3 was launched. Five minutes after liftoff, the parachute deployed, and the rocket went offcourse. The second A-3 was launched without a parachute. This one crashed as well. Von Braun decided to wait for the blustery winter weather to calm before sending the last A-3 aloft, but to no avail. The third rocket also went awry.

Disappointed, von Braun returned to Kummersdorf determined to learn everything he could about gyroscopes and flight stabilization. He and his team set about correcting the stability problem and some of the other snags in the A-2 and A-3 in a new test rocket they called the A-5. This rocket would test new stabilization systems devised by von Braun himself. Meanwhile, a much more ambitious project was also afoot: the A-4.

CUTAWAY VIEW OF THE GERMAN A–4 ROCKET

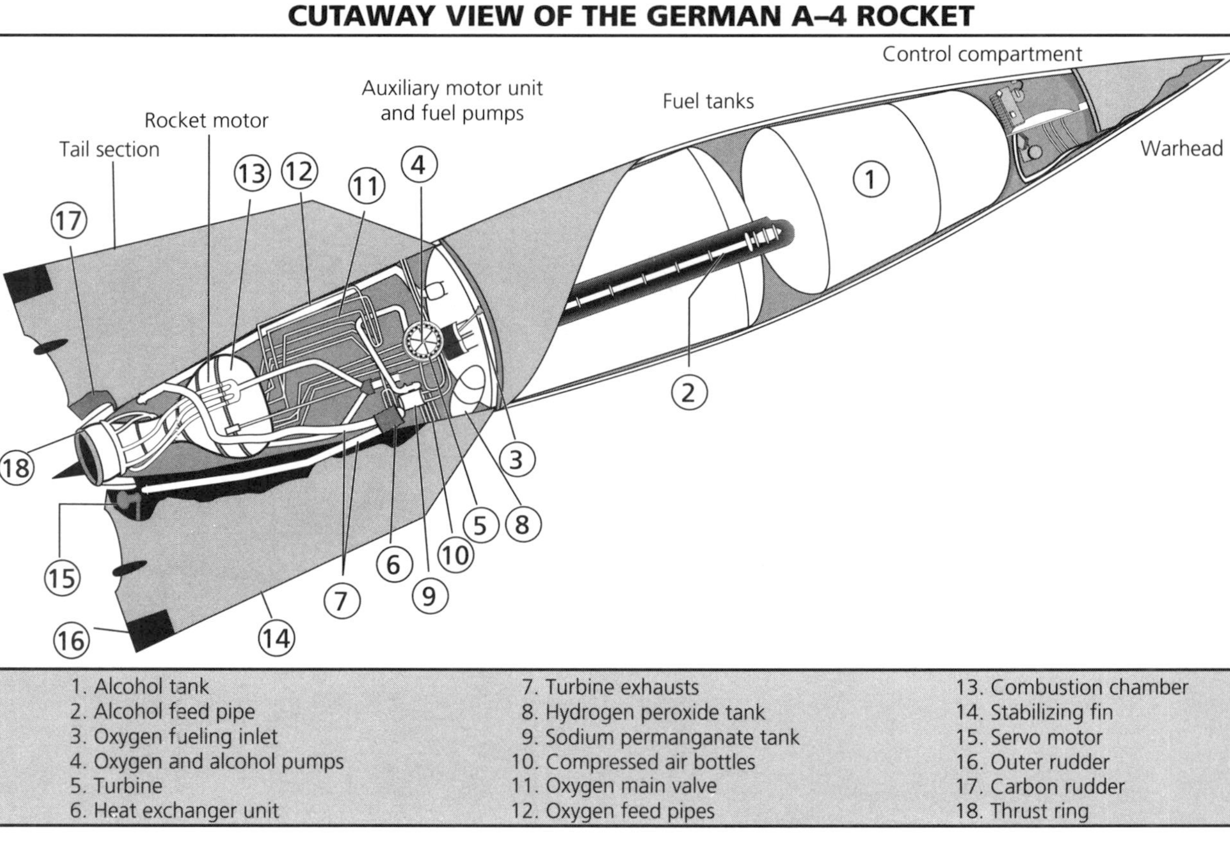

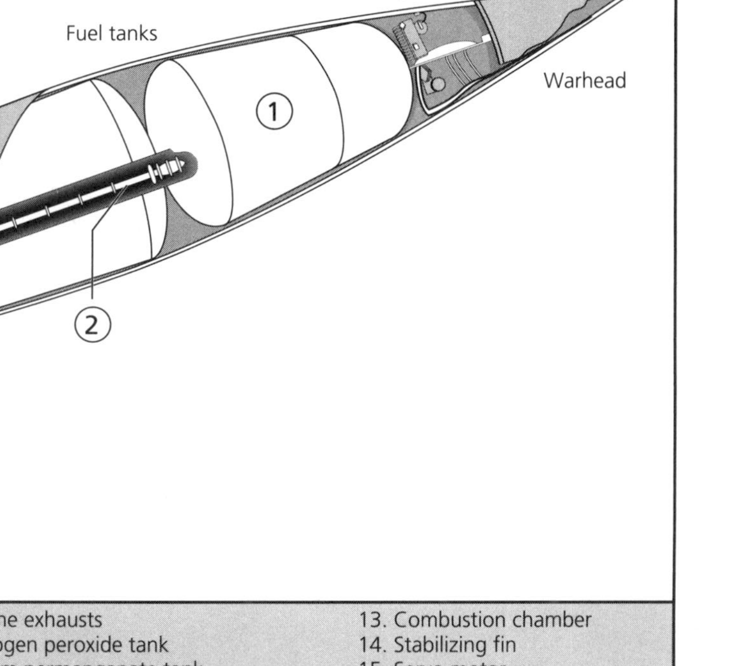

The A-4 was the first rocket to be built to specific performance specifications; earlier rockets were built and then tested to see how well the new design performed. A few days after General von Fritsch's visit in March 1936, Dornberger had sat down with Riedel and von Braun to outline his plans. They were developing artillery, he reminded them, for use in battle, and if they didn't come up with something useful, funding would dry up. He set out specifications for the next rocket, the A-4: It must have twice the range of the Paris gun, a total of 260 kilometers (about 162 miles); it must be capable of carrying a warhead weighing 1 metric ton (2,204.6 pounds); it could deviate only 2 to 3 meters (6.5 to 10 feet) from its target; it must be transportable by rail or roadway to any point within German boundaries (limiting its length and diameter to sizes that would pass through tunnels and go around curves in roads and railways).

One by one the members of the team faced and met the challenges. The team had grown between 1936 and 1939, with Dornberger and von Braun making several important additions to round out its technological strengths. Among them was Walter Thiel, a tense, ambitious, and brilliant engineer who combed his blond hair slicked straight back and wore horn-rimmed glasses. Thiel worked hard and was scrupulously conscientious. With his addition to the team, von Braun no longer had first-line responsibility for engine design, although he and Dornberger continued to brainstorm ideas with Thiel. By improving the mix of fuel after injection—thereby improving the overall efficiency of the rocket motor—Thiel succeeded in increasing the all-important exhaust velocity by 5 percent.

Dornberger suggested recessing the injectors, which worked well, and Thiel began clustering the injectors to increase power. Von Braun contributed the idea of placing 18 injectors in two concentric circles at the front end of a large combustion chamber. Another colleague suggested a cooling method for the throat of the nozzle, which tended to heat up and burn through with the passage of the hot exhaust gases. The new method involved using a film of alcohol to cool the metal in that area. These innovations combined to produce a truly reliable liquid-fuel rocket engine for the first time.

Most of the team moved to Peenemünde in May 1937, while Thiel remained behind at Kummersdorf until 1940 to continue his work on the engine.

But there was still more work to be done. The team needed to improve the fuel pump, and the specifications seemed impossible. It needed to be light in weight, fast reacting, simple in construction, and capable of delivering high volumes of fuel. Von Braun frequently made business trips flying himself in a Messerschmitt Typhoon. He could be anywhere in Central Europe within an hour or two, and he liked to visit manufacturers to obtain help with problems like this one. The answer this time was a fire-fighting pump, found at a pump factory, off the shelf and already in production.

Team members also found they could power the pump using high-strength hydrogen peroxide combined with a permanganate solution to produce a high-temperature, high-pressure steam to turn a turbine.

The aerodynamic problems did not respond immediately to this kind of brainstorming, however. To try out the stability of different fin arrangements and shapes at supersonic speeds, Dornberger had had the foresight to negotiate for the construction of a wind tunnel at Peenemünde, but it was one of the last buildings to be completed. Rudolph Hermann, a specialist in aerodynamics, was brought in from the Technical University at Aachen to consult on both the wind tunnel construction and the rocket's aerodynamic design. However, the final design decisions for the A-5 and, later, the A-4 were reached on the basis of tests made not in a wind tunnel, but by dropping test flight vehicles from Heinkel He-111 bombers at 20,000 feet. By the time the test vehicles had fallen to about 3,000 feet (11,090 meters) above the ground, they had reached supersonic speeds, and their behavior could be observed and filmed. Finally, the design engineers settled on an automatic pilot that was stabilized by a gyroscope, which in turn controlled air vanes on fins and jet vanes in the rocket exhaust.

During these days, von Braun spent evenings with his old friend, Hanna Reitsch, who had taken advanced gliding lessons with him at Grunau in 1932 and had since become Germany's foremost female aviator. She often performed test flights for the Luftwaffe

Adolf Hitler (hatless, at the center of the photo), who established himself as dictator in Germany in 1933, visited Kummersdorf in 1939 to see the army's advances in rocketry. (Archives, U.S. Space and Rocket Center, Huntsville, Alabama)

at Peenemünde West, and when she did, she and von Braun would meet for dinner and conversation. He enjoyed her zest for life and relished comparing notes about flying, which they both loved.

But the politics of building rockets proved to be far more complicated than either the technology or von Braun's social life. Both von Braun and Dornberger recognized that a project the magnitude of the A-4 development would not survive without the support of Hitler himself. So they were very pleased when, on March 23, 1939, Hitler came to visit.

Hitler did not, however, go to the new rocket center at Peenemünde, but opted for a look at rocketry closer to the capital, at Kummersdorf. Although people often have talked about "Hitler's rockets," Germany's leader was not a great fan of rocketry or most technologically advanced weapons. He felt that jet planes moved too fast for combat, and he deplored tanks as cowardly. But Dornberger and von Braun knew that, more and more in Germany, no project could flourish without Hitler's support. And whatever he opposed was as good as finished.

When Hitler arrived with his entourage, he seemed preoccupied, his thin lips pressed together beneath his little black mustache. Dornberger and von Braun began with the test stands, always a stunning display. But the 300-kilogram (660-pound) motor made no impression. Neither did the big 1,000-kilogram (2,200-pound) motor. Then von Braun gave a briefing on the A-5, using a cut-away model. As usual, he was at his best, careful not to insult or bore the Führer with too much detail, yet offering enough to whet his well-known interest in intricate machinery.

What kind of payload could this missile carry? Hitler wanted to know.

Von Braun explained that the A-5, a test vehicle, was not designed to carry any payload, but that larger rockets, based on this research, could be built to carry substantial military warheads.

How long would it take to develop such a vehicle? For this question von Braun deferred to Dornberger's military expertise.

"With our present level of effort, and with our present budgetary support, it will take considerable time," Dornberger responded carefully. Maybe, he thought, Hitler would see fit to increase the support.

But Hitler only nodded curtly.

Even the finale, a static test of the huge 1,500-kilogram (3,300-pound) motor, belching out its thunderous roar, didn't seem to move him. Later, over his usual mixed vegetables and glass of mineral water, he concluded, *"Es war doch gewaltig!"* ("Well, it was grand.") To the rocketeers, this odd comment, though apparently a sort of praise, seemed unenthusiastic and anticlimactic. Hitler, it seemed, was not overly impressed.

On September 1, 1939 German armies invaded Poland. Two days later, on September 3, Britain and France declared war on Germany, and World War II had begun. By 1942 Germany's opponents would be joined by the United States and the Soviet Union, which, with Britain, France, and China, became known as the "Allies."

On September 19 Hitler announced in a speech that Germany was working on a "new secret weapon," referring to the rockets, and for a brief time General von Brauchitsch gave the rocket makers the highest possible priority. Prior to the invasion of Poland, construction at Peenemünde, though nearly completed, had slowed almost to a standstill. The Luftwaffe had threatened to withdraw funding, and it looked like plans for the rocket center would fall through completely. But that autumn Albert Speer, the inspector general of construction for the Nazis, was placed in charge of completing Peenemünde. A favorite of Hitler's, Speer later confessed that he liked working with the young rocket team and von Braun, then 27, whom he described as "purposeful, a man realistically at home in the future." The work in rocketry, he said, "exerted a strange fascination upon me. It was like the planning of a miracle."

By the fall of 1939 most of the rocket team, with the exception of Thiel, had transferred their work from Kummersdorf to Peenemünde. Living quarters were completed, and the wind tunnel was almost ready. But the rocketeers were stunned to hear suddenly, as the year rolled to an end, that funding had been cut back again. Triumphant from an easy victory in Poland, Hitler had grown confident that the "secret weapon" would be unnecessary and had changed his mind. It was the first of a series of roller-coaster funding cutbacks and infusions.

Luckily for the future of rocket development, Speer stuck by the team, despite Hitler's directive. As Speer admitted after the war, "By tacit agreement with Army Ordnance Office, I continued to build the Peenemünde installations without its approval—a liberty probably no one but myself could have taken."

On June 13, 1942 the first A-4 was ready to fly. Armament chiefs from three branches of the armed services arrived to witness the test, along with Albert Speer. Dornberger, von Braun, and the team, all equally tense, gathered with the dignitaries to watch.

The rocket stood four stories high, commanding sober attention, a huge and improbable giant. As the fuel tanks were filled, wisps of vapor whirled into the crisp air.

The moment came. The beast seemed to tremble and then falter. Then with a deafening roar, the unleashed giant rose slowly upward. It seemed to pause for a moment, as if standing on the sword of its flame, and then shot upward, all 12 tons, with a great howl, into the low-hanging clouds. Von Braun's expressive face broke out in a beaming grin.

A minute and a half later another great howl told the startled observers that the rocket was falling back to Earth—quite nearby—plunging to the ground only half a mile (.8 kilometers) away.

In fact, the guidance system had failed—the stability system had not kept it from turning into the wind—but the great deed had succeeded: The mammoth beast had gotten off the ground. Corrections were made to the guidance system, and a second test was made on August 16, but again the rocket flew erratically. More work still had to be done.

Finally, on October 3, 1942, everything seemed ready. This time Dornberger watched from behind the protective brick parapet surrounding the roof of the Measurement House at Peenemünde, far enough from the launch site to be safe, but close enough to see.

"Would the launching be successful this time?" Dornberger found himself thinking. "Had we really discovered the cause of failure of the last two attempts . . .?" Ten years had passed since the team members had started out on this long path. Would they find success at last?

Countdown. Ignition. And the big rocket lifted straight up toward the sky as members of the team shrieked with delight.

"It looked like a fiery sword going into the sky," team member Krafft Ehricke would recall years later. "And then came this enormous roar. The whole sky seemed to vibrate. This kind of unearthly roaring was something human beings had never heard."

The A-4 arced through the sky, reaching up toward the fringes of space to an altitude of 60 miles (96 kilometers). The rocket builders were standing, they recognized, on the precipice of a new age; like Magellan and Columbus, they had sighted vast new worlds.

"This is the key to the universe," Dornberger is said to have exclaimed with great excitement. "This is the first day of the space age!"

Even Dornberger, as von Braun liked to tell it, had become infected, despite his military priorities, with the enthusiasm for space that his engineers all privately harbored.

In fact, they had produced a 12-ton missile that had traveled a distance of 120 miles (192 kilometers), under automatic control, landing within 1.5 miles (2.4 kilometers) of its target. "We have thus proved that it is quite possible," Dornberger told the team in a congratulatory speech, "to build piloted missiles or aircraft to fly at supersonic speed, given the right form and suitable propulsion. Our self-steering rocket has reached heights never touched by any man-made machine." In military terms, they had broken the world altitude record established at 24 miles (40 km) by the Paris Gun.

"We have invaded space with our rocket," Dornberger continued, "and for the first time—mark this well—have used space as a bridge between two points on Earth; we have proved rocket propulsion practicable for space travel. To land, sea, and air may now be added infinite empty space as an area of future transportation, that of space travel . . . So long as the war lasts, our most urgent task can only be the rapid perfecting of the rocket as a weapon. The development of possibilities we cannot yet envisage will be a peacetime task."

On December 22, 1942, Speer convinced Hitler to sign orders authorizing the mass manufacture of the rocket, which Hitler began to see as the potential answer to all his military woes. However, in March 1943, a new blow was delivered to the A-4 development. The always-erratic Hitler had dreamed, he said, that the rocket would never be operational against England, and he removed support from the project.

Nonetheless, on July 7, 1943 Speer invited Dornberger and von Braun to visit Hitler in his headquarters at the Führer's request. It had been four years since their last meeting. Germany had achieved no quick victory and the war was going badly. After a wait of several hours, finally the doors opened and a herald stepped forward with the Nazi salute and a cry: *"Der Führer!"* But there was something eerie about the leader when he entered the room. As Dornberger would later write:

> I was shocked at the change in Hitler. A voluminous black cape covered his bowed, hunched shoulders and bent back. He wore a field-gray tunic and black trousers. He looked a tired man. Only the eyes retained their life. Staring from a face grown unhealthily pallid from living in huts and shelters, they seemed to be all pupils.

But von Braun was confident, filled with his usual boyish enthusiasm. Assistants from Peenemünde showed a film of the great A-4 rocket launch, the huge missile rising from its pad and disappearing into the clouds. It was the first rocket launch Hitler had ever witnessed.

He suddenly grew animated, jumping up, asking questions. He was won over. Hitler promoted Dornberger to general and conferred the honorary title of "professor" on von Braun. He even apologized to Dornberger for not having believed in him.

"The A-4 is a measure that can decide the war," he confided later to Speer. "And what encouragement to the home front when we attack the English with it! This is the decisive weapon of the war."

Hitler's reaction was another stunning turnaround. But in some ways this new enthusiasm was worse than Hitler's prior disbelief in rockets. Now he was counting too heavily on the A-4 for powers it did not have.

In fact, it would not be the weapon to reverse the tide, the great thundering hope from the skies. The decision to pour resources into the A-4 rocket had come too late. For one thing, the A-4 was an experimental rocket, still a long way from being ready for production.

And yet another strike against the A-4's progress was about to descend, from out of the blue.

CHAPTER 5 NOTES

p. 39 "attack of acute generosity." Dornberger, *V-2* (New York: Viking Press, 1954), p. 41.

p. 46 "With our present level of effort . . ." Dornberger, p. 65.

p. 46 "Es war doch gewaltig!" Dornberger, p. 66

p. 47 "new secret weapon" Quoted in an excerpt from the November 1939 *Astronautics* journal in Frederick I. Ordway III and Mitchell Sharpe, *The Rocket Team* (New York: Thomas Y. Crowell, 1979), p. 104.

p. 47 "purposeful . . ." Albert Speer, *Inside the Third Reich* (New York: Macmillan, 1970), p. 366.

p. 47 "exerted a strange fascination . . ." Speer, *Inside the Third Reich,* p. 366.

p. 47 "By tacit agreement . . ." Speer, p. 366.

p. 48 "Would the launching be successful . . ." Dornberger, *V-2* (New York: Viking Press, 1954), p. 3.

p. 48 "It looked like a fiery sword . . ." Krafft Ehricke, in an interview aired in "Spaceflight," a multipart documentary shown on public television in 1986.

p. 49 "This is the key to the universe . . ." As remembered by Ehricke, in "Spaceflight" interview.

p. 49 "We have thus proved . . ." Dornberger, "The First V-2," in Arthur C. Clarke, *The Coming of the Space Age* (New York: Meredith Press, 1967), p. 30.

p. 49 "We have invaded space . . ." Dornberger in Clarke, p. 30.

p. 50 "I was shocked . . ." Dornberger, p. 101.

p. 50 "The A-4 . . ." Speer, p. 366.

6

PEENEMÜNDE UNDER FIRE: THE LAST DAYS

The moon was full on the night of August 17, 1943. It beamed out over the waters of the North Sea along the coast of Norfolk County on the easternmost edge of the British Isles like a great searchlight. Destroy "at the first favorable opportunity," the orders to British Bomber Command had read. And now the hour had come.

Wing Commander J. H. Searby took off at 9:50 P.M. from the Norfolk coast, flying southeast toward the Baltic Sea with his six-man crew in their Lancaster. Behind them followed 597 other Royal Air Force (RAF) Lancaster bombers and Halifax heavy bombers. It would be a long trip, and dangerous, since it was too far for the usual protective fighter planes to fly alongside as watchdogs. On this mission they were on their own. Sixty crews had not returned from a similar Allied bombing mission earlier that same day. This mission must be critical, Searby knew, or the RAF would never risk so many men at such unfavorable odds.

A flight of small Mosquito bomber planes had gone ahead, flying over Peenemünde without pausing as they streaked toward Berlin to feint an attack on the capital. The ruse worked. German defense aircraft followed them, drawn away as Searby and his bombers headed toward their mark. The pathfinder planes flew over the little island spit, dropping colored flares to signal the targets, Searby leading the way. As they zeroed in, finally the German antiaircraft defenses kicked in. They had grown wise to the diversionary tactics, but not before the British bombers were on their mark—and their mark was Peenemünde.

In Peenemünde that night, Dornberger had put out his after-dinner cigar and left the Hearth Room early. The conversation with aviator Hanna Reitsch, their guest that evening, had naturally drifted to flying. He had no real interest in the subject; it had been a hot, humid, trying day; and he was tired. As he made his way to the guest house where he usually stayed when he came up from his office in Berlin, he heard an early-warning siren. Peenemünde was usually secure—with good camouflage, nightly blackouts, and night fighter planes and antiaircraft guns at the ready. But that night the moon was full. He quickly phoned air defense headquarters to check on the warning. The reply was uncertain: Allied aircraft were massing in the central Baltic, but the status was unclear. Bombers often flew overhead en route to Berlin. This was probably just another foray of the kind that had become almost commonplace in recent days. Dornberger went to bed.

Von Braun, by this time the technical director of the Peenemünde rocket center, stayed a little longer to finish his conversation with the others. Then he walked Hanna Reitsch to her car. She would head over to Peenemünde West, where she was scheduled to test-pilot an Me163 rocket-powered plane the next day. He bid her goodnight and good luck and strolled to his own quarters, a dormitory where several bachelors lived.

Ernst Steinhoff, director of rocket guidance and control, who was also with the group that night, headed home to his family in the residential settlement, where many of the engineers and their families lived on the Baltic. Dornberger was jolted awake in the middle of night by the shattering of window glass as it sprayed across his bed and the sound of antiaircraft fire. He bolted from bed, grabbed his slippers, threw on his uniform tunic over his pajamas, and raced out. When he arrived at the air raid shelter, von Braun and between 50 and 100 other people were already there. Dornberger took charge immediately. He phoned the chief air-raid warden and received confirmation: They were in the midst of a full-scale attack. He set von Braun to work organizing a salvage party at the construction bureau. Every scrap of paper and machinery that could be saved had to be removed from the burning buildings.

Von Braun and others ran out into the inferno, surrounded by helpers covered with grime, sweat, and soot from the explosions. They dodged chunks of flaming debris and jumped gaping potholes as they made their way to the building that contained everything they had been working on, von Braun's life's work for the past 13 years.

His secretary, Hannelore Bannasch, later wrote in her diary about that night: "Everywhere I look it's fire. Fire, everywhere fire—horrible beauty!"

She and others rushed to the building that contained von Braun's office. The roof had already collapsed, but the stairway was still passable. What once were office doors now were completely unrecognizable. They fumbled their way to the room that was once his office, pressing tightly against the one remaining wall. The rest of the room had collapsed. They grabbed a safe and the files, stacks of them—as much as they could carry—and raced down the stairs, trip after trip, stashing them in the safe. Von Braun and some of the men threw equipment and files from the windows, while Bannasch gathered the papers into the safe outside as the fires raged around them. Finally the job was done and the most important documents had been rescued.

When at last the gunfire and bombing blasts grew quiet and the roar of airplane engines stilled, only the crackling flames filled the night. The British bombers had hit, and they had hit reasonably well. They also had paid a price. When Searby led the way back to England, 40 planes—and with them their crews totaling 240 men—were missing from the formation.

The next morning von Braun flew a plane overhead to take a look from the air. The place looked devastated, like the cratered surface of the Moon. Dornberger assessed the damage on the ground. The streets of Peenemünde were strewn with rubble, buildings blackened by blasts and steel girders twisted like pretzels. But not all was lost. Von Braun and his crew had saved the all-important blueprints. The test stands, wind tunnel, and Measurement House all stood unscathed. Of the 735 people killed in the attack, most were construction laborers—Russian prisoners of war and Polish slave laborers. Among the 178 German technicians killed, however, was the indispensable engine designer Walter Thiel, recently

The Royal Air Force attacked Peenemünde on the night of August 17, 1943, causing extensive damage to the rocket development facility. (Archives, U.S. Space and Rocket Center, Huntsville, Alabama)

moved to Peenemünde, whose house was hit while he and his family slept.

But the bombing had a devastating, though less direct, effect on the center and its work. As a result of the raid, it was obvious that Peenemünde was no longer a safe place to amass all the aspects of rocket development. Facilities were moved quickly to more dispersed locations. The valve lab was moved to Ankham, 19 miles (30 kilometers) to the south, and again, to Friedland, 42 miles (68 kilometers) from Peenemünde. The materials testing lab was moved to a vacant warehouse at Sadelkow, 43 miles (70 kilometers) south. By September the ground support equipment team—which was working on a launcher and a transporter—moved to two abandoned railway tunnels near Bonn, 358 miles (575 kilometers) to the southwest. And by November the supersonic wind tunnel was disassembled and shipped to the Bavarian Alps.

Whereas once the managers could visit all the shops under their jurisdiction every day, traveling by bicycle, now their charges were scattered all over the map of Germany.

Most telling of all, test launches were moved to Blizna, Poland, for the moment beyond the range of Allied bombers, and production was moved out of Peenemünde completely. As a result, Dornberger, who had been engaged in a subtle power struggle to maintain control of the A-4's development and production, stood to lose the battle.

Dornberger first had begun to lose control when Hitler had ordered the A-4 into production in December 1942 at the unreasonable rate of 2,000 per month. From the beginning Dornberger had planned to produce as many A-4s as possible at the preproduction plant he had built at Peenemünde South. But on January 8, 1943, Speer thwarted that plan by appointing a production engineer, Gerhard Dengenkolb, who took production out of Dornberger's hands.

In any case the A-4 was not really ready for use in the field, much less for this pell-mell production schedule. No one had even thought about building ground support equipment or methods of deployment in the field, and developing the necessary equipment and systems would take more than a year. Also, the A-4 was an experimental rocket with a lot of "bugs," or imperfections, still uncorrected. Each A-4 up to this point had been individually hand-built, and each one had 90,000 parts. Before the end of the war more than 60,000 design changes would be made *after* it was placed in mass production.

Following Hitler's visit with von Braun and Dornberger in July 1943, when the Führer suddenly began to see the A-4 as his "miracle weapon," the Peenemünde team was really placed in a bind. They were now expected—after months, even years, of cutbacks and delays—to produce a unique scientific instrument ready for production under wartime conditions. Throughout Germany Allied bombers now strafed industrial complexes and destroyed suppliers. The Allied armies advanced closer daily. And shortages made every step of production more difficult.

Another sinister political squeeze had begun to play in the picture. Even before von Braun and Dornberger's meeting with

Hitler, Heinrich Himmler, the chief of the SS, had appeared one day in April 1943 at Peenemünde. Dornberger thought the visit was odd. SS security was responsible for guarding the rocket facilities, but beyond that, the SS ordinarily had no jurisdiction on this joint army and air force installation.

But Himmler had been polite and had listened well as he looked through his pince-nez at the rocket test stands and the development facilities. When he left he said quietly, "I am extremely interested in your work. I may be able to help you. I will come again alone . . ."

Although the man had a reputation for extending his power through hard-line tactics, Dornberger didn't worry at first. Then Colonel Leo Zanssen, a friend who had been a station commander at Peenemünde for years and a steady and reliable career officer, was suddenly dismissed. Himmler himself had inexplicably brought charges against him. According to other friends, rumors among the SS officers blamed Dornberger for the delays in A-4 development.

Himmler came back, on June 29, driving himself in a little armored car. On this occasion he commandeered the attention of von Braun, Dornberger, and several other Peenemünde staff members for a five-hour lecture after dinner on the purposes of the war, history, and philosophy. The next day he saw two test launches of the A-4. The first failed, but the second flew beautifully. Himmler offered to put in a good word with Hitler for the rocket staff's work and left.

From the earliest development stages, one of Hitler's great concerns was that secrecy about the new weapon must be maintained. Now, after the raid on Peenemünde, the matter seemed all the more urgent, and for this, Himmler placed SS brigade general Hans Kammler in charge of rocket construction during the summer of 1943. Now, for the first time, the SS had made inroads into army territory. Kammler was also the SS officer Dornberger trusted least. Known to be devious, meddlesome, and even vicious, he undermined the team spirit and played Peenemünde workers against each other, even commenting once to Dornberger that von Braun was "too young, too childish, too supercilious and arrogant to be technical director."

Kammler had access to concentration camp laborers in his capacity as chief of SS construction, and he had a plan: Make use of prisoner-slave labor and no news of the "secret vengeance weapon" could possibly leak out.

An assembly plant was established at Mittelwerke, an underground factory tunneled into the Harz Mountains in southern Germany. Production of subassemblies and components began there in August 1943, and the first missile rolled off the assembly line that winter. On January 27, 1944, for the first time, a Mittelwerke missile was fired up for testing. It was an abysmal failure. Furthermore, the projected production was way behind schedule. Of the 650 planned to come off the assembly line in January, only 50 made it. Obviously the Mittelwerke production process still had a long way to go. Some 5,000 contractors and subcontractors in industry provided parts and systems to the Mittelwerke production initially, but as German industry suffered from Allied bombings, more and more of the work was moved to the dank underground tunnels of Mittelwerke.

There, and in other factories, Himmler's concentration camp inmates—totaling more than 30,000—were forced to work under incredibly inhumane conditions. By 1944 over 10,000 prisoners were at work in Mittelwerke alone, many of them making a forced march daily at 4:00 A.M. from the camp at Nordhausen. From another camp at Dora, the prisoners marched through a tunnel connected directly to the factory. Sanitary conditions were abominable, and both disease and starvation were widespread. Thousands died at their work as a result of their treatment, their corpses piled up unburied and rotting by the time Allied troops later discovered the camps.

During these years most of the rocket scientists, including von Braun, knew about the conditions under which the slave laborers worked. Later von Braun remarked, "The working conditions there were absolutely horrible. I saw the Mittelwerke several times, once while these prisoners were blasting new tunnels in there in what was a pretty hellish environment." He also told writer Arthur C. Clarke that he did not thoroughly inform himself about the conditions at Mittelwerke, "But I suspected it, and in my position I could have found out. I didn't, and I despise myself for it."

Heinous persecution and wholesale slaughter of Jewish citizens had already been going on for years under Hitler's regime. During the same years the rocket team was struggling for funding to keep their program alive, the Nazis stepped up their persecution of Jewish citizens, as well as other groups considered "unworthy," including gypsies, homosexuals and certain Catholics. Auschwitz was established as a concentration camp in 1940, after Poland's defeat, and by 1941 the first Jews and other victims began to arrive at what had by then become designated an extermination camp. Between 1 million and 4 million people died there in gas ovens and by other methods during the next four years.

What could von Braun have done about any of this? Probably very little. Generally, those who crossed Hitler and his henchmen paid with their lives and brought about no change. But many other Germans refused to cooperate and fled the country. Granted, leaving Germany in 1944 might have been difficult and certainly dangerous for the leader of the A-4 rocket development team—strategically he was very valuable to Hitler by this time. But it would not have been impossible. Von Braun, however, was wedded to his vision.

He later claimed that press censorship by the Nazi regime made the atrocities much less visible from within the Third Reich than from the outside. "I never realized the depth of the abyss of Hitler's regime until very late and particularly after the war, when all those terrible abuses were first published," he later wrote. Not seeing, or not looking, was an irreversible act for which many would hold him responsible throughout his life.

Meanwhile, the SS net continued to close around the work at Peenemünde. In February 1944 Himmler called von Braun in for a conference. No doubt, Himmler proposed, Peenemünde's technical director was fed up by this time with army red tape and bureaucracy. Things could go so much smoother if Himmler were in charge, instead of Dornberger. Why not throw in his lot with the SS? Himmler suggested to von Braun. Then progress on the A-4 could really take off, with Himmler personally clearing away the bothersome political obstructions.

Taken aback, von Braun answered carefully but firmly that no one could hope to find a more excellent administrator than

Dornberger. Any slowdowns encountered by the A-4 were caused by technical difficulties, not red tape, he assured the SS chief. He tried to conclude the meeting gracefully, and as he left, Himmler seemed at least polite. Von Braun returned to Peenemünde and threw himself back into his work, putting the strange meeting out of his mind.

At 3:00 on the morning of March 13, 1944, von Braun was wakened in his quarters by a curt knock at the door. Three Gestapo agents told him brusquely to dress and come with them to the Polizei Präsidium in Stettin. They had orders, they said, to bring him in for "safekeeping."

Von Braun had no idea what this "safekeeping" could mean—safekeeping from what, and why? He protested that surely there must be some mistake. But, in the end, it became clear he had no choice.

"I languished in the SS jail in Stettin for two solid weeks," he would later recall, "without the slightest information from the authorities as to the reasons for my arrest."

Finally he was brought, without explanation, before what appeared to be an SS tribunal. He was accused, they told him, of having said that he never intended the A-4 to be used as a weapon and that he had designed it for use in space travel. In fact, just a few nights before his arrest von Braun had been talking at a party about the rocket's potential for exploring space, as he often did. Yet the attitude was common among those working at Peenemünde. But that wasn't all, the SS continued. He was planning to escape in his airplane to England, with important data about rocket building. Von Braun would later remark to a *Life* magazine reporter, ". . . Himmler arrested me and accused me of keeping the plane gassed up so I could fly off to England with the V-2 [modified A-4] secrets, which was pretty absurd."

But at the time, the accusation struck him as chilling. He *was* in the habit of keeping his plane gassed up, for his frequent trips around Germany. If the SS wanted to interpret this fact as treason, how could he defend himself?

At that moment, Major General Dornberger marched in and presented the SS officers with a document, which they read immediately. The next thing von Braun knew, he was released and striding out of the police station with Dornberger. To von Braun it seemed like a miracle.

In fact, Dornberger's miracle was the result of two solid weeks of frantic work. Von Braun was not the only one arrested in the March 13 sweep in Peenemünde. The Gestapo also had scooped up Klaus Riedel and Hermann Gröttrup.

Dornberger had heard the news with a shock the next day, after being called to Hitler's headquarters in Berchtesgaden (in the Bavarian Alps in Southeastern Germany), where he had orders to meet with Field Marshal Wilhelm Keitel of the Armed Forces High Command.

"I could not believe my ears," Dornberger would later recall. "Von Braun, my best man, with whom I had worked in the closest collaboration for over ten years and whom I believed I knew better than anyone, whose whole soul and energy, whose indefatigable toil by day and by night, were devoted to the A-4, arrested for sabotage! It was incredible." The arrests of Riedel and Gröttrup seemed no less insane.

He appealed to Keitel. The matter was beyond him, Keitel explained, in Himmler's hands. If he tried to step in, the army's last liaison with Hitler would be gone, the only protection against the SS's rapidly widening control would be rubbed out. Suddenly Dornberger began to see the smoking gun. So that was it. Himmler was making his move for complete control of the A-4.

He asked to see Himmler. He was refused. But he did get a chance to see SS General Hans Kaltenbrunner at the SS Security Office. These three men, Dornberger asserted firmly, were vital to the A-4 program. Progress could not proceed without them. In the process, Dornberger was informed (probably to intimidate him) that the SS also had a fat file on *him.* Speer would later claim that von Braun and his colleagues were released through *his* intervention, rather than Dornberger's. Quite likely the efforts of both men had an effect. In any case, the three rocket men were released provisionally for three months, and at the end of that time the adjournment was renewed.

During the years 1943 to 1944, von Braun continued working on new designs—including a rocket he called A-7 and two others that he called A-9 (also known as the A-4b) and A-10. The A-9 and A-10 were intended to work together as a multistage rocket—an idea

The V-2 rocket. (Archives, U.S. Space and Rocket Center, Huntsville, Alabama)

anticipated by both Tsiolkovsky and Oberth—with the A-9 boosted by an A-10, producing a projected range of some 3,200 miles (5,120 kilometers). For the first time, practical engineering was done on the idea, and this multistage rocket, if completed, would have been able to strike New York City.

While Germany never developed that capacity, the A-4 became a weapon of terror in London and Antwerp during the years from 1944 to 1945. The first attacks on September 8, 1944 took England completely by surprise. The last of the dreaded attacks by the Luftwaffe's V-1 "buzz-bomb" (*Vergeltungswaffe 1,* "Vengeance

Weapon #1") had dwindled off during the summer, and British officials had declared the "war of terror" at an end at last.

But the sigh of relief was premature. The armed A-4, or V-2 (*Vergeltungswaffe 2,* "Vengeance Weapon #2"), as Hitler's propagandists called it, was in some ways worse than the V-1. It seemed to come from nowhere, and it plunged straight to Earth, striking all the way to the basements of buildings, where the din of its explosion rocked everything for miles and whole rows of houses were taken out. By the end of war, the Germans had manufactured between 5,789 and 6,915 V-2s, and 3,225 landed on targets in England, France, Belgium, and elsewhere. The V-2s killed over 2,700 people in England, seriously wounding many more. In Antwerp, a single rocket caused 271 deaths, and many thousands of homes and buildings were destroyed.

Perhaps, when he first became a civilian employee of the army, von Braun had not imagined his rockets would really be used as weapons. But no such case could be made for him later, when "Hitler's rockets" became Nazi "vengeance weapons." Wernher von Braun was a pragmatic man who had seen a way to achieve his goals and followed the path with focus and discipline. When the time came that his progress might be hampered because he was not a member of the Nazi party, he joined the Nazi party, and, with Peenemünde under SS control, he also became an SS officer, according to FBI records collected later.

Von Braun had his own agenda—never losing sight of his original vision of building space rockets, which he talked about openly and often (at least once too often). But he also never seemed to regret the use made of his work, in contrast to many American scientists, such as Robert Oppenheimer, who admitted to deep anguish when the atomic bomb they had created was used against citizens in the cities of Hiroshima and Nagasaki. For this apparent lack of conscience, von Braun came in for criticism throughout his later life. But his personal ethic called for singleness of purpose and discipline, for performing his best, and for keeping commitments. He may have regretted that he had not informed himself about the workers at Mittelwerke, but on the subject of rocket development, he felt he had performed only with honor.

Should von Braun, then, have acted differently? The SS was right about one thing—he could have used his gassed-up plane to fly away. Thousands of others did leave, including Albert Einstein and von Braun's old friend Willy Ley. But to von Braun, the thought was "absurd." His attitude let him in for criticism worldwide as a proponent of "science without conscience" or, as a popular satirical song of the 1960s put it, "'Once the rockets are up, who cares *where* they come down/That's not my department,' says Wernher von Braun."

In the end, the V-2 did not vanquish the English; if anything, the huge rockets strengthened their resolve, and they came back fighting harder than ever. Soon the Allies had Germany in a great pincer, the British and the Americans coming from the West, the Soviets advancing from the East. France was retaken, then Belgium. The Allied bombers strafed any areas from which they thought the V-2s could be launched, and soon the Germans were forced to retreat from almost all areas that were within the V-2's 200-mile (124-km) range to England. From Peenemünde, the sounds of the Soviet guns could be heard in the all-too-near distance. By the end of 1944, the handwriting was on the wall: Germany would not win the war. The time for decisions had come.

CHAPTER 6 NOTES

p. 54 "Everywhere I look . . ." Quoted in Frederick I. Ordway III and Mitchell Sharpe, *The Rocket Team* (New York: Thomas Y. Crowell, 1979), p. 130.

p. 57 "I am extremely interested . . ." Dornberger, p. 181.

p. 57 "too young . . ." Quoted in James McGovern, *Crossbow and Overcast* (New York: William Morrow, 1964), p. 48.

p. 58 "The working conditions . . ." From an interview taped in 1976, shortly before his death, and aired in "Spaceflight," a multipart documentary shown on public television in 1986.

p. 58 "But I suspected it . . ." Arthur C. Clarke, *Astounding Days* (New York: Bantam Books, 1990) p. 184.

p. 59 "I never realized . . ." In a letter to R. W. Reid, quoted in Reid, *Tongues of Conscience: Weapons Research and the Scientists' Dilemma* (New York: Walker and Company, 1969) p. 105.

p. 60 "I languished in the SS jail . . ." Quoted in McGovern, *Crossbow and Overcast,* p. 52.

p. 60 ". . . Himmler arrested me . . ." *Life* magazine, November 18, 1957, p. 139.

p. 61 "I could not believe . . ." Dornberger, *V-2* (New York: Viking Press, 1954), p. 201.

p. 64 "Once the rockets are up . . ." Tom Lehrer, "Wernher von Braun."

7

NARROW ESCAPE: OPERATION PAPERCLIP

As the days marched into 1945, Soviet guns approached nearer and nearer to Peenemünde. Polish refugees began to stream across the island of Usedom, instead of crossing the Oder River at its mouth, as a shortcut in their efforts to escape the Soviet troops pressing them from the east. Old people pushing wheelbarrows filled with their belongings trudged hopelessly through the snow. Young women and children huddled together against the cold. The tales they told of rape, murder, and pillage at the hands of the Soviets did nothing to brighten the Peenemünde team's vision of the future. Orders came down from the German army command to stand and fight, but everyone knew that rocket scientists would make poor opponents for trained Soviet soldiers and their weapons. However, the SS had arrested and shot several engineers who had been heard saying so and hung their bodies as examples from trees along the busy roadways. Signs with large letters made the point: I WAS TOO COWARDLY TO DEFEND THE HOMELAND.

The Allied forces in Europe, meanwhile, were pushing the German troops out of the areas from which they had been launching V-2s. The Peenemünde team was increasingly pressed to increase the V-2's range, without slowing the production schedule. Shortages and inferior quality of component parts slowed the work markedly, and von Braun sent his brother Magnus to Mittelwerke to see what he could do to expedite production. In Peenemünde, transportation, supplies, and communications became increasingly critical issues, while ration stamps, black-marketeering, and complaints became rampant. The evidence of the damage done to German industry mounted daily.

By August 1944 Himmler had appointed Hans Kammler to a new post, placing him in charge of all A-4 development and manufacture. Although Albert Speer protested that this move infringed on his territory, the overlapping authority had remained in place, and the iron fist of the SS clenched all the more firmly around Peenemünde and its crews. Dornberger was shunted off to Berlin for another project. In part to divert Allied intelligence from suspecting that Peenemünde was still operational, the installation was renamed *Elektromechanische Werke* (Electromechanical Works) and became a civilian, state-owned industrial firm with a general manager, Paul Storch, effectively taking Dornberger's place. Although Storch was in charge, he deferred in all technical matters to von Braun, who remained on the scene as technical director. Von Braun, however, had neither the power nor the influence to evacuate anyone or anything from Peenemünde, and the SS appeared to have no thoughts of evacuation.

Then on January 31, 1945 von Braun called in his section chiefs and department directors.

"Kammler has just ordered the relocation of all the most important defense projects into central Germany," he said in a strained voice. "This is an order, not a proposal."

Now that the order had come from Kammler, somehow it seemed suspect. Why was he massing all the defense projects in central Germany? In a way, it seemed like a reasonable plan: Mittelwerke was there, and if the Allies were pressing in, Germany could then consolidate forces in an effort to stand them off. But von Braun suspected Kammler had other motives in mind, perhaps wanting to gather some pawns to play in his favor when the time came to surrender to the Allies. And von Braun knew he didn't want himself and his men to be part of such a plan. Another possibility was that the SS sought to eliminate everyone who might be useful to the enemy; they could be walking into a deadly trap. However, at the moment, he seemed to have no choice.

At 8:00 the morning of February 3, von Braun held another meeting. Slated for the move were 4,325 people. "But we will go as an organization," von Braun asserted. "This is important. We will carry our administration and structure straight across Germany. This will not be a rout."

A hundred trucks were loaded with equipment and files. Barges were loaded on the river, and train cars were packed.

They moved out by night for the 250-mile (400-kilometer) journey south, with the first trainload departing on February 17, 1945, less than three weeks after Kammler's orders had been received. More than 500 technicians and their families were on board.

By mid-March only a skeleton crew was left at Peenemünde. The test stands and launch sites stood empty. Altogether, 264 developmental launchings of the A-4 had been fired between June 13, 1942 and February 19, 1945. An era had come to an end.

As his engineers and technicians began arriving at their destination in the Nordhausen-Bleicherode area, von Braun pressed ahead with developmental work on more advanced rockets, since the A-4 now seemed to be a lost cause. He set up the experimentation sites in the underground Mittelwerke factory, or wherever he could in unused garages and empty buildings.

The tension and ceaseless activity were exhausting, not only for von Braun, but for his crew. On a road trip, his driver fell asleep at the wheel, crashing the car and fracturing von Braun's arm and shoulder.

Then more bad news. On March 19 Hitler, recognizing that defeat was near, ordered the SS and the army to destroy everything of possible value to the approaching enemy forces. By March 21 von Braun had argued his way out of the hospital and began planning to save the 65,000 V-2 drawings and tons of blueprints and documents from what became known as Hitler's "scorched earth" policy—even though ignoring the Führer's orders could mean big trouble with the SS.

He asked Dieter Huzel, his personal assistant, and Bernhard Tessman, who had saved him days before from the wrecked automobile, to gather up all the V-2 technical documents, load them on trucks, and find a hiding place.

Then on April 1 the danger to the Peenemünde rocket specialists became especially ominous. Kammler had decided he would retreat with 500 of von Braun's key scientists and engineers to the

Bavarian Alps. They could continue their research at an old army camp at Oberammergau, and they would be protected by a special SS detachment.

Kammler claimed that his plan was connected to a scheme of Hitler's to regroup with key SS divisions in a place 400 miles (640 kilometers) south, known as the Alpine Redoubt. From there they would make a stand and emerge victorious. But von Braun felt sure that Kammler was actually executing his own plan to play a deadly game of chess using the Peenemünde men as pieces. Of course, disobeying Kammler's orders could still bring immediate execution, so they went to the Alps, with von Braun hoping he would have a chance to save his team. Dornberger, meanwhile, was able to position himself in a village close by with a detachment of army soldiers who were under his command.

On April 4 von Braun arrived in Oberammergau. The camp in which his men were billeted was surrounded by barbed-wire fence, and as each day passed, he worried about what would happen as the scene played out. By April 10 work had stopped at Mittelwerke, and the 4,500 technicians who remained there were dispersed to the surrounding villages. By April 12 the Nordhausen area was in the hands of American troops.

For months, as the war drew to a close, independent intelligence teams from each of the Allied countries had been jockeying to secure German weaponry, engineering designs and brain power. Preparing for postwar animosities, each country hoped to get away with the most and gain the greatest advantage, once the regions of Germany opened up. Now the Mittelwerke factory's V-2s and V-2 parts had fallen neatly into the hands of Americans, who immediately set about locating more and tracking down the documents Huzel and Tessman had hidden. The soldiers managed to dig them out of the mine they were stashed in mere hours before the British moved in.

As the German rocketeers watched the Allied forces press in on them, they debated about which nation would be most advantageous to surrender to, if given any choice. They settled on the Americans. When later asked why, one member of the rocket team explained, "We despise the French, we are mortally afraid of the Soviets, we don't believe the British can afford us, so that leaves the Americans."

But von Braun and all the key people were still 400 miles away, in grave danger and in no position to surrender to anyone.

Then suddenly Kammler summoned von Braun to a meeting. The conversation between them in a local tavern was strange. Kammler asked how von Braun's men were, how they took the transfer, and whether they had resumed their all-important work for the future. He then explained that he was leaving to tend to other duties. Final victory, he assured von Braun, was still attainable. That was all, dismissed. Von Braun never saw him again.

The next day von Braun took advantage of Kammler's departure to convince the major who had been left in charge that concentrating all these valuable minds in one place was not a good idea. A single hit could destroy the men who had created the Third Reich's greatest technical triumph. Better they should be dispersed in the local villages. His persuasive, telling arguments won the day, and he and Ernst Steinhoff, the Peenemünde director of guidance and control, succeeded in moving the men out from behind the sinister barbed-wire fence.

About April 25, von Braun was roused from his sleep by a German army soldier with a Red Cross armband. He was whisked away to a hospital, where his arm was reset, and then transported to the village of Oberjoch near the Austrian Tyrol. There, at a hotel, he found Dornberger, his brother Magnus, and several others of his rocket team colleagues. Dornberger had gathered as many of his rocket experts as he could, hoping to protect them if necessary against the SS with the 100 men he had at his disposal.

By May 1 word arrived that Hitler had committed suicide. The end, they knew, was finally at hand.

The following day von Braun sent Magnus off on a mission, through the snowy woods, on his bicycle. Magnus von Braun's instructions were to find the Americans and tell them where Dornberger, von Braun, and the rest were located. But when he finally encountered an antitank unit on the lonely country road, they didn't believe him at first. They took Magnus in for questioning. Later that afternoon, when Magnus brought his rocket colleagues in under safe conduct from the U.S. Army, and the Americans met Wernher von Braun, the group was still deeply skeptical that this energetic 33-year-old man could be the famous

Von Braun (center, with cast) at the time of his surrender to the United States Army. (Archives, U.S. Space and Rocket Center, Huntsville, Alabama)

German rocket genius. This man, they thought, was too young, too fat, too jovial.

But quickly the word was passed to a group of American agents who had been charged for months with finding this select group of Germans whose names had been placed on what was known as the "Black List." The British, French and Soviets had put together much the same list, but it was the American army that had undramatically come into posession of these important human "spoils of war."

The Americans were elated at their find. Colonel Holger N. "Ludy" Toftoy, U.S. Army Ordnance Corps, moved quickly to negotiate to bring as many of the German scientists and engineers as he could to the United States. Whichever nation succeeded in enlisting their help, he reasoned, could begin where the German experts had left off, instead of "reinventing the wheel." Although the V-2 had not won the war for Germany, if it had been developed earlier, it might have. And if an atomic warhead (which Germany had not developed) had been added, the V-2 would have been devastating.

In a semicovert operation known as "Overcast" (renamed "Operation Paperclip" in 1946), Toftoy enlisted von Braun's help in choosing the best group of irreplaceable scientists and engineers from his team.

Von Braun and the others underwent extensive interrogations in Garmisch and Paris, all the while being propositioned by the Soviets to walk out on the Americans and come over to their rocket program. Only one of them did—Hermann Gröttrup, who had been arrested by the SS with von Braun and Riedel in 1944. According to his wife, who later published her diary of their years with the Soviets, he lived to regret the choice. The Soviets promised the German rocketeers that they would be allowed to work in Germany and then virtually kidnapped them and their families and whisked them away to the Soviet Union. They were never trusted and spent much of their time at test sites in desolate Siberia. As von Braun remarked in 1957, "The Russians let them see nothing, touch nothing on the production end, learn nothing from developing experience, just had them write reports until they were drained dry. Then the Russians went on by themselves." By 1957 not one of the German rocket scientists and technicians who went to the Soviet Union was still there. They all either had died or had been sent back to East Germany.

The British had also succeeded in gathering up V-2s and V-2 parts, and in June 1945 von Braun proceeded to London "on loan" to help the British with questions they had about their V-2 rockets. Several other rocket specialists, including Walter Dornberger, also were part of this "loan." While the German scientists were in England, the British tried to entice many of them to give up the contracts they had signed with the Americans and stay to help them with their rocket programs. The deal was attractive—London was a lot closer to home, and the British promised to put them up in posh hotel accommodations. Von Braun and the rest were cordial but didn't take them up on their offer, returning to Paris on schedule.

Dornberger, however, did not come back. The British wanted to keep him to stand trial on charges as a war criminal for his role in the use of the V-2 weapons. Although Dornberger protested and von Braun contributed testimony, Britain kept him for two years, finally releasing him in recognition that he, in fact, had no jurisdiction over the firing of the V-2s. Dornberger finally arrived in the United States in 1947 and accepted a position with Bell Laboratories, where he completed his career.

Back in Paris, Toftoy finalized arrangements for von Braun and 114 others from the team to leave for the United States. In mid-September, just over four months after surrendering in the Bavarian Alps, Wernher von Braun's plane took off for what would become his new home, where he would begin a new chapter in rocket history.

CHAPTER 7 NOTES

p. 67 "Kammler has just . . ." Dieter K. Huzel, *Peenemünde to Canavaral* (Englewood Cliffs, NJ: Prentice-Hall, 1962), p. 133.

p. 67 "But we will go . . ." Huzel, p. 136.

p. 69 "We despise the French . . ." James McGovern, *Crossbow and Overcast* (New York: William Morrow, 1964), p. 199.

p. 72 "The Russians let them . . ." Von Braun in an interview with George Barrett, "Visit With a Prophet of the Space Age," *The New York Times Magazine,* October 20, 1957, p. 86.

8

FORT BLISS AND WHITE SANDS: V-2s IN THE AMERICAN SOUTHWEST

In the spring of 1945, trainloads of strange cargoes began wending their way into the desolate desert regions of southern New Mexico to a place called White Sands. The cargo contained fully assembled, V-2 rockets, as well as parts and materials for building them.

Not long afterward 115 German engineers, technicians, scientists, and mechanics followed to a nearby army installation in Fort Bliss, Texas. Fort Bliss was in the far northwest corner of Texas, just north of El Paso, where the Rio Grande separates the United States from Mexico and a jog in the border of New Mexico accommodates the north Texas border. It would become home to these men who had become a strange sort of war booty.

As a result of military orders from the supreme Allied commander, General Dwight D. Eisenhower, Wernher von Braun arrived in the United States by air in mid-September 1945, along with several other top members of his rocket team. He reported first to Fort Standish in Boston, Massachusetts. From there he proceeded by boat to Fort Strong, on New York's Long Island, where he and six other German rocket scientists underwent long hours of interrogation. To pass the time when the Army Intelligence Service wasn't questioning them, they played games of Monopoly far into the night.

At this point many of von Braun's colleagues were sent to work at the Aberdeen Proving Grounds in Maryland, sorting and interpreting 14 tons of notes and documents that they had produced while developing the German rockets. The task was arduous and tedious. Eisenhower's original orders read, "Upon completion of this duty, the civilians named below will return to their proper station in this theater." Most of them, however, never did return to their homeland, except to visit.

On October 3, 1945 the U.S. Army Ordnance Corps activated a sub-office for rocket development by creating a technical unit located at Fort Bliss, Texas and placed Major James P. Hamill in command. From Long Island, accompanied by Hamill, Wernher von Braun set out across the country by rail for Texas. On the way, for his own protection, Hamill encouraged von Braun not to mingle with other passengers. Memories of the war and lost loved ones were all too recent in the minds of Americans, and his fellow passengers would hardly have welcomed the news that they were traveling with a former Nazi.

In Fort Bliss, von Braun was soon reunited with most of his hand-picked German rocketeers. Their job: to put together and test the V-2 rockets and rocket parts that had arrived from Germany, as part of a project named "Hermes," shared by General Electric and the U.S. Army. While the German rocket specialists hoped they would be able to continue the work in space rocketry that they had begun, at that time, no one in the United States was thinking in those terms yet.

For the most part, the five years at Fort Bliss were years of isolation for the German rocket team. They were not allowed opportunities to meet with American engineers to talk shop; nor were they allowed to attend scientific conferences or read journals in their field. Their lives were carefully structured, controlled, and isolated. On weekends, small groups of five or six were allowed to go into town, in the company of an army escort, to shop, have dinner at a restaurant, or maybe see a motion picture.

In those early years, some segments of the U.S. government kept the rocket specialists under constant surveillance. Each specialist was assigned an army "custodian," who kept a security dossier on him and was responsible for his whereabouts and activities at all times.

In addition, the Germans received low pay, had an uncertain future, and underwent separation for many months from their

Von Braun playing ball at Fort Bliss, Texas. (Archives, U. S. Space and Rocket Center, Huntsville, Alabama)

families. Colonel (later Major General) Toftoy, Major Hamill, and von Braun somehow convinced the team that all these problems were worth putting up with for the sake of an opportunity, nebulous and uncertain as it must have seemed. As rocket engineer Ernst Stuhlinger later recalled, "After all, we came here to help develop rockets and we wanted to do that."

Toftoy once remarked that von Braun "held his team together through many years of adversity," which he was able to do because the rocket specialists knew what they had already accomplished as a team. They knew that von Braun could bring out the best in them—and they counted on having the opportunity to excel together again.

Hamill and Toftoy personally went a long way toward making the transition from Germany to the U.S. work. As one participant in Operation Paperclip later wrote, "General Toftoy, Colonel Hamill and their staff did not act as bosses, but as leaders and friends."

Their work, meanwhile, kept the scientists busy. And it eventually attracted the attention of journalists. From October 1945 to

May 1950 the team built and fired V-2s at White Sands, New Mexico. As the science editor of *Time* magazine wrote at the time in his book on the subject, "A rocket shoot at White Sands Proving Ground is more than interesting, more than beautiful, more than exciting. It is inspiring in a way that is equaled by few sights on earth."

In 1947 von Braun returned to Germany briefly to marry his 18-year-old cousin, Maria Louise von Quistorp. The wedding took place at Landshut, Bavaria on March 1. Later that month the newlyweds returned to the United States, accompanied by von Braun's parents, whose estate in Silesia had been confiscated by the East German government.

The first of Maria and Wernher von Braun's children, Iris Careen, was born in 1948. Her sister, Margrit Cecile, would be born four years later, in 1952, followed by a brother, Peter Constantine, born in 1960. Wernher von Braun, along with all the rest of the German scientists and engineers of the rocket team, was putting down roots in America.

But many American citizens felt a real concern about bringing former enemies into the country to work on sensitive military projects and paying them a salary out of taxpayers' money. Many of these scientists had been members of the Nazi party; some, including von Braun, also had accepted commissions in Hitler's dreaded SS. At the time, some of these facts were glossed over, even in the top-secret FBI files; in fact, changes were made in their histories to reduce Americans' objections to their staying in the United States.

In 1980 journalist Linda Hunt compared documents she obtained through the Freedom of Information Act. According to her article, published in *The Bulletin of the Atomic Scientists,* and presented in a television documentary, Hunt proved that the files had been altered. In von Braun's case, the original security evaluation written in 1947 read:

> Based on available records, subject is not a war criminal. He was an SS officer but no information is available to indicate that he was an ardent Nazi. Subject is regarded as a potential security threat by the Military Governor, Office of Military Government for the U.S. A complete background investigation could not be obtained because subject was evacuated from the Russian Zone of Germany.

With the documents that could have been used to do a more thorough evaluation out of reach, no further inquiry was planned. Von Braun's position as an SS officer could be explained away because Peenemünde had come under SS rule at the end of the war, although this background, if generally known, would have made many people uncomfortable. But the phrase "a potential security threat" could have damaged his chances for U.S. citizenship, and without achieving that critical step, the country could not put many of von Braun's talents to use. The report was later altered, according to Hunt's evidence, to read:

> Further investigation of Subject is not feasible due to the fact that his former place of residence is in the Russian Zone where U.S. investigations are not possible. No derogatory information is available on the subject individual except NSDAP records, which indicate that he was a member of the [Nazi] Party from 1 May 1937 and was also a Major in the SS, which appears to have been an honorary commission. The extent of his Party participation cannot be determined in this Theater. Like the majority of members, he may have been a mere opportunist. Subject has been in the United States more than two years and if, within this period, his conduct has been exemplary and he has committed no acts adverse to the interest of the United States, it is the opinion of the Military Governor . . . that he may not constitute a security threat to the United States.

The doctored evaluation, as it turns out, was more accurate than the original. Without question, not only was von Braun not a security threat, but as an "opportunist" who wanted to build rockets, he was a pronounced asset to the government of the United States, which wanted them built.

In one case, the U.S. Justice Department did later take action. Arthur Rudolph was a key engineer in rocket engine design at Peenemünde and later—as program manager of the Saturn V rocket—became a central figure in the American space program. But during the war years in Germany, according to the Justice Department, he had held a responsible position at Nordhausen, the inhumanely run concentration camp that provided prisoner-slave laborers for V-2 production at Mittelwerke. Prior to his time at Nordhausen and after it, Rudolph had worked at von Braun's side for most of 40 years, in Kummersdorf when they had shared bachelors' quarters and visions of space, in Peenemünde, at White

Sands, and later, in Huntsville, Alabama. In 1983 Rudolph received word that he had to surrender his U.S. citizenship and leave the country or face charges of war crimes. Although he denied the charges, he left the United States for Germany at the age of 76.

As the years passed, von Braun grew weary of the constant surveillance and questioning about his "Nazi background," although he understood the reasons for it. In the 1950s he once remarked to an acquaintance, "That's the only thing I don't like about America," as he motioned to an FBI agent who was following the car he was traveling in. "They follow me wherever I go."

The original stated purpose of the German team's work was to train Americans, both civil and military, in the assembly, checkout, and launch of the large rockets they had designed. In the process, they would use the big V-2 rockets and their parts to gather data on the physical environment and radiation of the upper atmosphere.

On May 10, 1946 the first A-4 was launched to carry scientific instruments into the upper atmosphere. Several more followed that year, and, as always, the team had many failures as well as successes. The highest altitude attained by a U.S. A-4 was 132 miles (nearly 670,000 feet). (By contrast, the experimental high-flying rocket plane known as the X-15 could fly only about half that high, and weather balloons reached altitudes only one-fourth that high.)

Demands for experimental space aboard the A-4s from government research agencies, the military, universities, and industry became so great that the A-4 Upper Atmosphere Research Panel was formed on January 16, 1947 to arbitrate requests and allocate space on flights. Between 1946 and 1951, the Hermes program, as it was known, launched 67 V-2s.

Another project developed during this period used specially adapted A-4s with modified nose cones. Called "Project Blossom" and sponsored by the U.S. Air Force Air Materiel Command and the Aero-Medical Laboratory, this program sent up canisters containing insect and plant life to study the effect of radiation on life forms at very high altitudes. Several rockets also carried mice and monkeys. In these earliest U.S. attempts to test the effects of ascent to extreme altitudes on living organisms, the nose cones were supposed to return to Earth via parachute. But the parachutes kept

By 1946 the former German rocket team was launching V-2 rockets in White Sands, New Mexico. This is the 1950 launch of a U.S. Hermes missile built from V-2 parts. (NASA Marshall Space Flight Center Archives)

fouling and the ejection mechanism was not perfected, so the recovery method was poor, and few primate test animals survived.

In a project promoted by Toftoy, the team at White Sands also developed the Bumper missile, of which eight were launched between 1947 and 1950. The Bumper was a two-stage rocket—the

A-4 was the first stage and a WAC-Corporal sounding rocket, weighing only 661.5 pounds, was the second stage. These rockets could go higher than the A-4 alone and could measure temperatures and cosmic radiation at much higher altitudes. Bumper No. 5, launched February 24, 1949, reached an altitude of 244 miles, a record at that time. It marked the first U.S. experience with large two-stage rockets.

In 1948 the United States government began dealing with a strange problem that had arisen with Operation Paperclip specialists. Because they had entered the country as wards of the army, they had no passports or visas. Although the original six months of their contract had long since expired, officially they had never entered the country. The State Department, reluctantly convinced that most of these men were not national security risks, agreed that they could begin proceedings toward citizenship. But citizenship could not be awarded until five years had passed following the date of legal entry. The State Department therefore came up with an ingenious plan. Interested specialists could take a streetcar trip from El Paso across the Mexican border to Ciudad Juárez, where they would visit the American embassy and obtain visas for entry to the United States. From there they would hop the streetcar again and return to El Paso. Port of embarkation: Ciudad Juárez. Port of debarkation: El Paso, Texas. Von Braun and his team had now officially arrived.

Despite their contributions, though, and increasing evidence that they might stay longer, by 1948 von Braun's team began to feel frustrated. They began to refer wryly to their presence in the United States as "Project Icebox," because they began to feel that their talents, especially as a team, had been kept on ice for three years, with no apparent hope of a thaw. They felt their work was moving too slowly and was underfinanced, and that the project was directed by an ill-defined policy. They were especially disappointed by the lack of team projects, the kind of work at which they had excelled at Peenemünde.

Then one day in 1950, Wernher von Braun gathered his team together to make an announcement. The White Sands testing ground had grown too small. (A misguided V-2 that had nearly caused an international incident by landing across the border in a cemetery in Ciudad Juárez had proved the point—even though the former German rocket team liked to joke that they were the only German detachment to attack Mexico from within their base in the United States.) The army had decided to establish a new testing site at Cape Canaveral, Florida, where rockets could be launched over the water, as they had in Peenemünde. Von Braun and his rocket team, he explained, were about to say good-bye to the American Southwest and head east, to a little town in northern Alabama. It was a town he thought they would like, and there they would continue their work in rocketry.

Few of the scientists knew at the time that they were about to enter into a love affair with Huntsville, Alabama and with American culture—about to become truly part of the suburban life of the 1950s, complete with outdoor barbecues, patios, and swimming pools.

Wernher von Braun was also about to embark upon a new phase of his life, one in which he would become a well-known public figure, his name a household word, and his visions of space the talk of the news media. He was about to become America's "practical prophet of space."

CHAPTER 8 NOTES

p. 75 "Upon completion of this duty . . ." From a reproduction of Eisenhower's orders in John Goodrum, *Wernher von Braun, Space Pioneer* (The Strode Publishers, 1969), p. 48.

p. 76 "held his team together . . ." Quoted by Clarence Lasby in *Project Paperclip* (New York: Atheneum, 1971), p. 124.

p. 77 "A rocket shoot . . ." Jonathan Norton Leonard in *Flight into Space* (New York: Random House, 1953), p. 9.

p. 77 "Based on available . . ." Linda Hunt, "U.S. Coverup of Nazi Scientists," *The Bulletin of the Atomic Scientists,* April 1985, Vol. 41, No. 4, p. 19.

p. 78 "Further investigation . . ." Hunt, p. 19.

p. 79 "That's the only thing . . ." Quoted by Charles Shows, in an interview with the authors, December 1993.

9

PROPHET OF THE SPACE AGE

Wernher von Braun didn't really set out to be a prophet or a public figure. He was above all a practical man who believed in getting things done the most efficient way possible. In an ideal world, this would have involved 100 percent secrecy for his work and all the money he needed, not unlike the way the result-oriented Communist government in the Soviet Union ran things in the 1950s. But von Braun understood that the price he paid for American freedom was American openness.

"When the Kremlin wants ballistic missiles," he once said wistfully, "it tells the scientist to meet the schedule and doesn't worry about public relations." But that wasn't the American way. "Here," he added, "we must have money and public support. Congressmen must believe in what we're doing, and they won't until the public believes in us." So Wernher von Braun began to take the case for space to the public.

When the German rocket team packed up for the move from White Sands to Huntsville, von Braun was 38. Even seven years later, when he was in his mid-40s, he still looked boyish, still had a broad, winning grin, and, as one reporter put it, "sometimes lets out a Falstaffian boff that reverberates through the corridors a little like the explosive echoes from one of his mighty Jupiters [rockets]." His enormous zest for life was catching, and he swept others—both those working with him and the public in general—up in his enthusiasm for his work. His early talent for languages came in handy now—although he always retained a Germanic staccato in his consonants, his English was loose, even slangy, always colloquial, and very American.

Von Braun began to use his talents as a speaker for space in the U.S. as early as 1950. (NASA Marshall Space Flight Center Archives)

Wernher von Braun knew how to communicate and he did it well, using the well-honed skills—first developed in his days with the VfR in Germany and later at Peenemünde—to promote his lifelong dream of sending humans into space.

He wrote articles. He made speeches. He produced plans for future spaceships, timetables for planetary expeditions, and blueprints for a space-based shipyard where space rockets could be produced. His ideas were far-ranging and ambitious. But his credentials and the solid proof his rockets represented kept people from putting him in the "crackpot" category. The respect he won in the United States was enormous.

He began by finishing a book he had been working on for years, *The Mars Project,* in which he outlined his vision of planetary exploration and space travel. However, every one of the 18 U.S. publishers he submitted it to turned it down—they thought it was too fantastic and improbable. It was finally published first in Germany in 1952, and then republished by a university press in the United States in 1953.

From March 22, 1952 to April 30, 1954 *Collier's* magazine published a series of articles that predicted some aspects of the future with uncanny accuracy. The articles came out in eight installments, written by members of a panel of experts brought together to discuss space realistically and to offer a blueprint for space exploration. The articles grew out of three symposiums on space sponsored by the American Museum-Hayden Planetarium in New York.

Collier's treated the articles as future science fact, not science fiction, and they were a conceptual tour de force. Von Braun's articles provided factual, detailed descriptions of spaceships and launch vehicles, based on work he had already done and was currently doing at the Redstone Arsenal in Huntsville. Chesley Bonestell, the highly talented space illustrator, provided many of the images for the articles. The impact was enormous, and the message that came across was, "All this is really possible, and not so very far off in the future."

Many obstacles to reaching space are purely imagined, von Braun liked to say. As he told one reporter:

> A rocket propulsion expert will tell you there's no sweat in solving propulsion problems but the medical problem is the barrier. A doctor will tell you space flight is medically possible but he doesn't think propulsion can be licked. Many of these obstacles will fade away. Look at the sound barrier and the heat barrier.

Always a space visionary, von Braun became director of Marshall Space Flight Center in 1960. (NASA Marshall Space Flight Center Archives)

His optimism was compelling, his wry sense of humor winning. "Right now," he added cannily, "we're up against the cash barrier, and that one doesn't always fade away so quickly."

Von Braun made as many as 50 speeches in one year when he was stumping for the army's missile program in the early years at Huntsville. He knew that getting the word out was an important part of the job and one of the keys to success.

But perhaps the most stunning result of all from the *Collier's* articles was the attention they gained from one of the great entertainment mavens of all time, Walt Disney. There are conflicting stories regarding who gave Disney one of von Braun's articles, or even whether it was one written by him. But as a result of seeing the material, Disney zeroed in on the subject of space as the theme for the "Tomorrowland" segment of his widely popular "Disneyland" TV show. He contacted Willy Ley (who had come to the U.S.

in 1935), to design the segment and Ley put him in touch with von Braun, Ernst Stuhlinger, and Heinz Haber. Together they brainstormed their way to putting together a stunningly visual series of programs, complete with rocket models and von Braun's narration. Von Braun worked with the Disney artists to produce a fantastic drawing of a spoke-wheeled space platform, with human beings stepping off into gravity-free space and shuttle rockets coming and going. It was probably thanks to the public exposure on "Disneyland" that von Braun became the only nonastronaut associated with space whose name became a household word.

Von Braun soon became so widely known that he finally had to get an unlisted phone number at home—especially after a man called one New Year's Eve asking for a ticket to the Moon. By 1957 he would get about 10 letters a day on space and traveling to the Moon, about half of them from kids wanting to know how to pursue a career in rocketry—to which von Braun would reply, as his own teacher had, to study math and physics heavily.

To promote his cause, he used his full arsenal of humor and boyish charm. He also could use the ammunition of fear persuasively. In a 1952 speech he urged that the United States build a "manned satellite to curb Russia's military ambitions."

By the late 1950s Congress frequently consulted von Braun, who provided extensive testimony at hearings, in addition to his other speaking engagements and duties at Huntsville.

And then suddenly what still seemed to many people like space fantasy became space history: A beeping ball that signaled its presence from beyond the atmosphere to every nation on Earth was launched. This tiny, spinning ball of metal was lobbed into space by a giant military power on the other side of the world.

CHAPTER 9 NOTES

p. 83 "When the Kremlin wants . . ." Von Braun in *Life* magazine interview, November 18, 1957, p. 136.

p. 83 "sometimes lets out a Falstaffian boff . . ." George Barrett, *The New York Times Magazine,* October 20, 1957, p. 86.

p. 85 "A rocket propulsion expert . . ." *Life* magazine interview, p. 136.

p. 87 "manned satellite . . ." Von Braun, quoted in *Life* magazine, p. 133.

10

HUNTSVILLE: THE STAGING GROUND

On October 4, 1957 the Union of Soviet Socialist Republics launched a 23-inch ball of polished metal weighing 184 pounds into orbit around Earth. The world would never be the same again. This little blinking ball was the first artificial satellite, and with its launch, the space age had begun.

Beeping and flashing across the night sky where all could see, this new object seemed at the same time magical and menacing. The Soviets called it *Sputnik,* and its presence became an extraordinary symbol of Soviet technical expertise and rocket power. Within a month they had launched a second *Sputnik,* this one weighing a whopping 1,120 pounds and carrying a dog aboard, again into Earth orbit. In the United States, the effect was galvanizing. With the launch of the *Sputniks,* an intense, 12-year "Space Race" against the Soviets began in the United States, and Wernher von Braun's dream of sending humans into space was the ultimate goal.

But the Soviets' launch of *Sputnik* had not surprised the U.S. government as much as many Americans thought. And Wernher von Braun was probably the least surprised of all. Two years earlier he had submitted plans to launch an American satellite based on the rocket technology already developed and tested at that time.

But while von Braun had been writing articles for *Collier's,* entertaining interviewers from *Life, This Week,* and other magazines, and working with Walt Disney on futuristic views of space travel, the U.S. government had made a strange decision. In 1955, it had decided to limit von Braun's team, working for the army at the Redstone Arsenal, to developing intermediate-range missiles, leaving the longer-range intercontinental ballistic missiles (ICBMs) to the air force and the project of launching a space satellite to the navy.

The Redstone crew had been busy during the last seven years developing two sturdy rockets, the Redstone and the Jupiter missiles. Despite the limitations, von Braun and "The Team," as he called the rocket crew, had modified one of their Jupiter rockets by adding another rocket on top. This two-stage rocket had flown at least 3,600 miles in a test and had reached an altitude of 600 miles—higher than *Sputnik's* top altitude, although not fast enough to enter orbit. But it could have.

In much the same way that he once had contended with interference from the German SS, von Braun now had to contend with interference from U.S. political factions. In an interview with *Life* magazine in early November 1957, von Braun remarked:

> We should declare a moratorium on obstructive inspections and monitoring by scientific committees and let the people building the Atlas, Titan, Thor, Jupiter and Polaris [rockets] work in peace. I just wish someone had the authority to tell me, "All right, we'll leave you alone for two years, but if you fail we're going to hang you."

Despite his complaints, the move von Braun and his colleagues had made to Huntsville was the best career opportunity of their lives. At first, though, it might not have seemed that way. Huntsville was a little cotton town in the South, its greatest claim to fame the champion milk-producing cow in America. When the army decided to revive the arsenal in the town, local residents weren't at all sure this was good news. Arsenals had come and gone before, and when they closed, the damage to the local economy was enormous. Little did its residents suspect that their town would soon become known as "Rocket City."

In 1950 Wernher and Maria von Braun and their two-year-old daughter, Iris Careen, had moved into a small house with a big, four-paneled picture window looking out on the neighborhood and the rolling Alabama countryside. By 1957 they were well settled.

At work, as usual, von Braun was organized, efficient, and practical. In 1957, he still drove a weatherbeaten Chevrolet and carried his lunch of apples and pears, which he ate while catching up on his reading. Usually he was the first to arrive at the closely guarded building, passing rows of combination-locked, safelike office doors on the way to his third-floor corner office. His desk was stacked with piles of documents stamped TOP SECRET—scores of production reports and memos about technical disasters and breakthroughs.

Always a hands-on manager, from his Kummersdorf days his colleagues liked to say that von Braun was never satisfied until his hands "felt the touch" of each piece of a rocket's mechanism.

"I've never felt my job was to sit in my office and think," he once said.

> Missile building is much like interior decorating. Once you decide to refurnish the living room you go shopping. But when you put it all together you may see in a flash it's a mistake—the draperies don't go with the slip covers. The same is true of missiles. Sometimes you can take one look and see something obviously wrong—not accessible perhaps, or too flimsy. The people who are working with it all day are too close to see it. That's why I go to the fabricating shop—I want to know what my baby will look like.

He liked to gauge success by the "looks and feel of the hardware."

Without question, the October 7 launch of *Sputnik* had sent the country into a turmoil, and Wernher von Braun immediately turned up in the calm eye of the hurricane that followed. The United States, he assured the many reporters who asked, did not lack the genius required to compete in the arena of rocket technology. No country, he said, could hold a monopoly on technical intelligence. As he told *New York Times* reporter George Barrett in an October 1957 interview:

> Take a group of people with the same general fund of knowledge, present them with a problem and you will get it done. The idea of a great genius sitting quietly in some corner dreaming up vast secrets is no longer real. So many things in modern science have become so wide in scope, so intricate, that more and more it takes groups of experts to do the work.

The same laws of physics applied everywhere in the universe, he liked to point out. All that was required was rolling up one's sleeves and getting down to work—teamwork.

But in the flurry following *Sputnik*'s launch, President Eisenhower decided not to go with von Braun's Redstone rocket. Instead, he decided to stick with the untried rocket being developed by the navy, the Vanguard.

Frustrated, even angry, von Braun and General John Medaris, commander of the Redstone Arsenal, went immediately to Secretary of Defense Neil McElroy to plead their case.

"Vanguard will never make it," von Braun said. "We have the hardware on the shelf. For God's sake, turn us loose and let us do something. We can put up a satellite in sixty days, Mr. McElroy! Just give us a green light and sixty days!"

In later years von Braun tried to put a less parochial cast on the frantic opinions he expressed at the time. Vanguard is, in von Braun's later words, "one of the most sophisticated and one of the finest pieces of engineering I have known . . ." However, he added, "Vanguard is a young racing horse, and one can expect it to be a bit capricious."

It was, to say the least, capricious the day the navy invited TV cameras and press to watch the historic liftoff. Tension, of course, was high, as Kurt R. Stehling, head of the Vanguard Propulsion team, relates in his account of the launch:

> T-0 seconds. The final fire switch was closed. The last second-stage umbilical cord dropped. The rocket engine began to show sparks and fire as the pyrotechnic igniter in its inside ignited the beginning of the oxygen and kerosene fumes. The time was 11:44:559 A.M. America's first satellite was about to take off.
>
> The engine started with a heart-rending, hoarse, whining moan like that of some antediluvian beast in birth pain. Flame filled the nozzle, dispiritedly at first, and then built up with a great crescendo to a tremendous howl, brilliantly white, streaked with black. The vehicle shook itself momentarily like a wet dog. Ice and snow fell off the sides. The banshee howl of the engine increased. The vehicle hesitatingly ripped itself loose from its iron womb and rose slowly. We rose up with it on our tiptoes.

At that moment, someone screamed, "Look out! Oh, God, no!"

Suddenly bright flames stabbed out from near the engine. The whole rocket seemed to pause, deathly still for a moment, and trembled. Then, as shocked observers looked on, instead of lifting powerfully off the launch pad, the big rocket began to topple, sinking downward into itself, breaking apart, colliding against the test stand, crashing to the ground. Even behind the protection of the two-foot concrete walls of the blockhouse and the six-inch-thick bulletproof glass, the deafening roar and the giant tremor jolted every observer. The launch pad area was flaming red with fire.

When finally the flames were doused, Stehling and his team could see out the windows into the steam and smoke. "The fire died down," Stehling would later write, "and we saw America's supposed response to the Russian 200-pound *Sputnik* satellite—our 4-pound 'grapefruit'—lying amid the scattered glowing debris and, unbelievably, still beeping away, unharmed."

Though the satellite, miraculously, had survived, the humiliation could not be overcome by anything but stoic courage. The moment of horror was over and could not be reversed. The navy team's attitude was, okay, time to clean up and get the next rocket ready.

The next shot, however, would not go to the navy. Concerned about the humiliation caused by the very public Vanguard disaster, President Eisenhower gave von Braun and his team the nod.

Von Braun and his team leapt into action, and by January 31, 1958—not much over his promised 60 days—the Jupiter-C rocket put together at Redstone sent a small satellite named *Explorer 1* zooming into space.

Only 15 years had passed since that blustery October day in 1942 when von Braun and his rocket team had stood on the shores of the Baltic Sea to watch the first successful V-2 rocket fly into the clouds. Now they recaptured that exhilarating moment once again—only this time the goal was unquestionably space.

By March 3, 1959 von Braun's team also succeeded in launching the *Pioneer* probe. It was the first satellite to orbit the Sun successfully and the premier performer of an exciting series of spacecraft that would explore the planets, sending back images and data.

Von Braun's visions of space travel and planetary exploration were beginning to come true at last.

Explorer 1, **perched atop a Jupiter C (modified Redstone) rocket prior to launch in January 1958.** (NASA)

CHAPTER 10 NOTES

p. 90 "We should declare a moratorium . . ." Von Braun in interview with Richard B. Stolley, *Life* magazine, November 18, 1957, p. 136.

p. 91 "I've never felt . . ." Interview with Stolley, pp. 134–135.

p. 91 "Take a group . . ." Quoted by George Barrett, *The New York Times Magazine,* October 20, 1957, p. 86.

p. 92 "Vanguard will never . . ." Quoted from Clayton Koppes, *JPL and the American Space Program* (New Haven, MA: Yale University, 1982), p. 83, by William E. Burrows, *Exploring Space: Voyages in the Solar System and Beyond* (New York: Random House, 1990), pp. 73–74.

p. 92 "one of the most sophisticated . . ." Erik Bergaust, *Reaching for the Stars* (New York: Doubleday, 1960), p. 28.

p. 92 "T-0 seconds." In "Vanguard," in *The Coming of the Space Age,* edited by Arthur C. Clarke (New York: Meredith Press, 1967), p. 19.

p. 94 "The fire died down . . ." Stehling, p. 19.

11

SPACE GAINS CENTER STAGE: POST-*SPUTNIK* FLURRY

Despite the *Explorer* success, the embarrassment of *Sputnik* would not go away. For one thing, while the U.S. satellite had served a scientifically useful purpose—detecting radiation belts around Earth (sometimes called Van Allen belts) that no one had known existed—*Explorer* was still just a tiny grapefruit compared to the big *Sputniks*. And everyone knew that it took massive power to put a heavy chunk of metal out into Earth orbit. The superior lift power of the Soviet Union's rockets translated into superior missile power if they were used as intercontinental missiles, aimed at another country instead of at the stars. Unlike the German Nazi government, which had wanted to squelch all talk of space, the Soviets were using space to tell to the world about their military might. In the Cold War rules of the day, the United States had to talk back. Clearly the U.S. would have to meet the challenge, and the obvious next step was to put a human into space. Von Braun and his team submitted a proposal they called "Project Adam" shortly after *Explorer's* launch in 1958. Its purpose was to send an astronaut into space. At about the same time, the National Advisory Commission for Aeronautics (NACA) proposed a similar project to put men into orbit around Earth. Both proposals were taken seriously.

None of this would have been considered even remotely by the United States in the 1950s, though, if the Soviet Union hadn't forced the issue. President Eisenhower had no interest in pursuing space exploration, but clearly he had no choice. In an effort to defuse the military impact of the Soviet Union's success, he reformed NACA in 1958 as the National Aeronautics and Space Administration (NASA), a civil agency that would be in charge of future space projects.

By the spring of 1959, a small group called the Space Task Group at Langley Research Center in Virginia had begun to talk about what would be involved in a manned lunar landing. It was the first time any of the group had really thought about it. Von Braun was there, from Huntsville, and Max Faget, a creative young Langley engineer. Maybe the astronauts should orbit the Moon, Faget suggested. Look down on it with binoculars as they flew by. Von Braun was impatient. They could just go in for a landing, he said. Already he had laid out in his mind the series of steps they would follow. An unmanned landing program called Surveyor was already planned.

"Max, you're completely overlooking all that we're going to learn in the Surveyor Program," von Braun said. By the time a manned landing was made, he surmised, the process would be old hat.

The first step in proceeding with any kind of manned space program was to test out the air force's Atlas rocket. On June 9, 1959 "Big Joe" lifted off from Cape Canaveral in Florida carrying a Mercury space capsule, the tiny little spacecraft that astronauts eventually would travel in. It reached a height of 160 miles, and then the Atlas spun around and sped back into the atmosphere at speeds up to 17,000 mph. From measuring devices attached to the capsule's titanium shingles, the scientists could find out how much heat this reentry would generate, and they could see how well the shingles held up. Similar test launches continued throughout the development of the manned space programs, revealing much about the various systems with each test.

Two months after the Big Joe launch, Eisenhower officially transferred Wernher von Braun (along with the rest of his team) to NASA, which had established the brand-new George C. Marshall Space Flight Center in Huntsville, right next door to the army's Redstone Arsenal.

Von Braun's strong point always was getting things done. He could bolt down great quantities of work in a single gulp. He could command a meeting of people in a way that generated dedication and enthusiasm, that led rather than brow-beat. Those who worked with him respected the wisdom of his experience and his ability to state his case coherently and cohesively.

"Von Braun," his former personal assistant, Dieter Huzel, once wrote, "didn't conduct a meeting, he led it with perception and incisiveness. He knew most problems at first hand, and those few that he didn't, he knew by instinct bred of long, intimate and successful experience in the field."

The Space Task Force at Langley would be responsible for developing the spacecraft, or capsule, while the Marshall team would develop the launch vehicle. (The Space Task Force moved to Houston, Texas in 1961, where it became the Manned Spaceflight Center and later the Johnson Spaceflight Center.) In 1960 von Braun became the director at Marshall, a position he would hold for the next 10 years.

For the two groups, the immediate goal was to send a human astronaut into space. For von Braun and his rocket-building team, the challenge became to build rockets capable of lifting spacecraft and human cargo there. For them it was the dream come true at last.

The Mercury program, the first of several increasingly more complex programs, was begun in November 1958. It was designed to send one person at a time orbiting around Earth. All did not go well, however. An empty Mercury capsule was launched atop an Atlas launcher in July 1960 to test the capsule's ability to reenter the atmosphere on a ballistic (free-fall) trajectory, but the rocket failed. A second test combining the Mercury capsule with the Redstone rocket also was unsuccessful due to rocket failure. No one wanted to send an astronaut into space without trying the effects of the journey on a test animal first. But the first test animal to go did not make the trip until January 31, 1961, when Ham the chimpanzee took the ride into space atop a Redstone rocket. Several things went wrong—errors during liftoff subjected him to an incredible 17 gs (17 times the effect of gravity), and the spacecraft leaked when it splashed into the ocean. But Ham seemed fine. (Ham lived to the age of 26 in a North Carolina zoo, where he died in 1983.) Ham's good health on his return to Earth encouraged the astronauts' confidence in the system.

Astronauts (from left) John Glenn, Scott Carpenter, and Gordon Cooper meet with von Braun (holding cigarette) during a visit to the Marshall Space Flight Center prior to their Mercury space flights. (NASA Marshall Space Flight Center Archives)

The first manned flight—like Ham's flight, a suborbital lob from a Redstone launcher—would not take place until May. The delays had eaten up time.

Then the news came in. On the morning of April 12, 1961, the Soviet Union announced that Cosmonaut Yuri Gagarin had made an orbital flight of one hour and 48 minutes. Everyone knew the Soviets had been close to a launch, but the length of the flight was stunning.

Alan Shepard, the first U.S. astronaut to enter space, would not make the trip until May 5. His flight consisted of a 15-minute hop in the capsule he had named *Liberty Bell 7.* Unlike Gagarin, however, he did not ascend into Earth orbit.

Once again the Soviets were too far ahead to catch easily—the Atlas rocket wasn't ready yet and, originally, all seven Mercury astronauts were scheduled to make *sub*orbital flights using the smaller Redstone launcher before trying to reach Earth orbit. Now that plan was scrapped. Only Shepard's flight in May and Gus Grissom's flight in July would be launched by the Redstone unit, and neither would reach orbit. Not until February 20, 1962, 11 months after Yuri Gagarin, would the first American complete an orbital flight in space: John Glenn aboard a Mercury capsule launched by an Atlas rocket.

In the same week as Gagarin's flight, President John F. Kennedy faced the low point of his administration—a group of Cuban patriots had stormed the Bay of Pigs in Cuba from the shores of Florida, with Kennedy's prior knowledge and approval. But when they were outmatched by 20,000 Cuban troops, Kennedy could not risk escalating the conflict into a war with Cuba's ally, the Soviet Union. He had to leave them unsupported.

Now the Soviet Union had beaten him at the space game again. Clearly he would have to make a move. Kennedy saw that he could make one of three choices: Quit the race entirely, remain second to the Soviets, or make a commitment to beat them to the Moon. The first two were not good options.

After much deliberation on ways and means and what was possible—including consultations with von Braun and others—Kennedy made his decision. With Shepard's first space flight safely completed, on May 25, 1961, in a speech to a joint session of Congress, John F. Kennedy proposed that the United States should establish as a goal a manned mission to the Moon within the decade.

CHAPTER 11 NOTES

p. 97 "Max, you're completely overlooking . . ." Quoted by Charles Murray and Catherine Bly Cox, *Apollo: The Race to the Moon* (New York: Simon and Schuster, 1989), p. 42, from an interview with Maxime Faget.

p. 97 "Von Braun didn't conduct . . ." Huzel, *Peenemünde to Canaveral,* p. 126.

12

TRIUMPH AND ACHIEVEMENT: SATURN SOARS AND THE EAGLE LANDS

When President Kennedy announced the U.S. commitment to send a man to the Moon within 10 years, everyone knew it would take a committed, focused, all-out effort. At that time the United States did not have rockets powerful enough even to put a man into orbit, much less send him beyond Earth orbit, far out into space.

That was von Braun's end of the job. On the surface it could be made to seem a puny task. All he had to create was a piece of machinery that could operate for under 18 minutes total: It would burn 6 million pounds of fuel in bursts lasting 2½ minutes and 6½ minutes, followed by 2-minute and 6-minute bursts. And that was it. But putting together a rocket that had the necessary power *and* control was no small matter.

There were, of course, many other complexities—the design of the main spacecraft (known as the Command Module), decisions about the design of the Lunar Module (a secondary spacecraft to transport astronauts from the Command Module orbiting the Moon to the lunar surface and back to the Command Module), and endless details involving the safe transport of astronauts through the unforgiving, airless expanse of space. So much about what might happen out there was not known.

But if von Braun and his team did not do their jobs well, those astronauts might never even get off the ground.

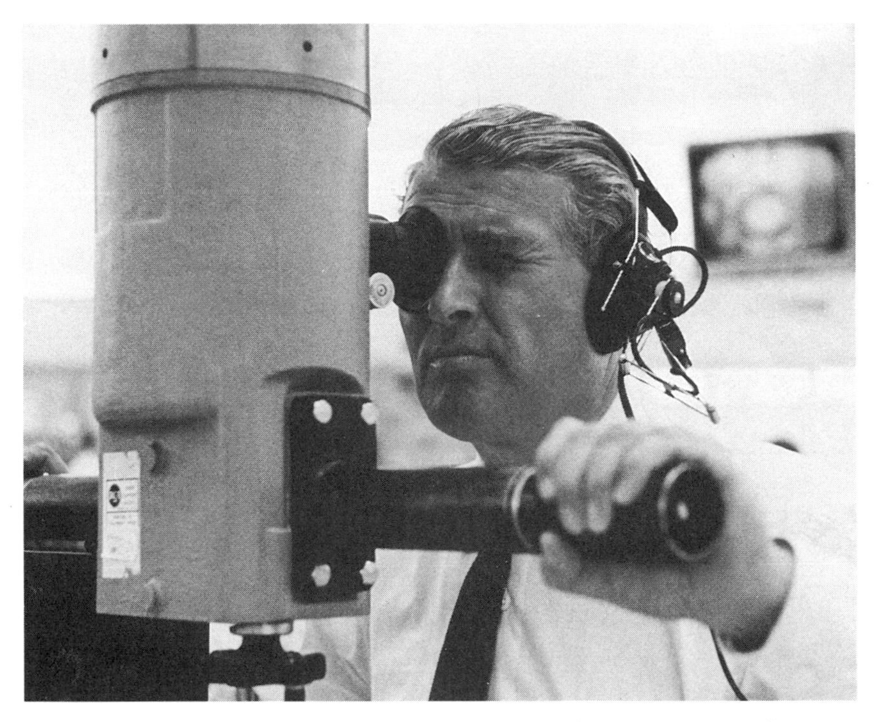

Von Braun uses a periscope to check on a rocket launch at Kennedy Space Center. (NASA Marshall Space Flight Center Archives)

The Apollo program would build on the lessons learned from Mercury and from a second program, Gemini, which sent two astronauts up together in a spacecraft. During the Gemini program, the astronauts completed many exercises that were necessary in planning a successful Moon landing, including docking with another spacecraft, extravehicular activities (EVAs), and transferring from one spacecraft to another.

For the rocket builders, plans for a Moon mission meant dovetailing their work developing rockets with that of other branches of the Apollo program. The biggest job, of course, would be sending a crew of astronauts, the Command Module and the Lunar Module to the Moon. For this job, von Braun and his team immediately began designing the giant Saturn V rocket. They began with several smaller vehicles that could work either together, as the Saturn V, or separately, for less daunting tasks than the Moon landing.

Von Braun gives a briefing to President Eisenhower on Saturn engine development during a visit to Marshall Space Flight Center in 1959. (NASA Marshall Space Flight Center Archives)

On October 27, 1961 the first Saturn vehicle was launched, the Saturn I. It was 162 feet tall and weighed 460 tons, "fully loaded" at liftoff. It traveled to an altitude of 85 miles and made a nearly flawless run.

The work was going well, and by 1962 the von Braun family had moved from their first Huntsville house on McClung Street to a new, trilevel home on a narrow, winding street called Big Cove Road, located on a shady hillside known as Monte Sano.

Meanwhile, the Gemini flights continued, now using the early Saturn rockets. The design concept was functioning reliably—it was safe to move on to perfecting the Saturn V.

From the beginning the design principle called for close analysis of test flight data. The guiding notion was the safety of the crew, and the team recognized early on that confidence in the rocket, especially considering its size and potential volatility, had to be based on the soundness of the engineering design. If a rocket flew perfectly three times, testing just to build confidence wasted resources and produced meaningless statistics. A rocket that has a three out of four success ratio still may blow up one time out of four. When you have people on board, those are not acceptable odds. The rocket had to work every time.

What the team created was an incomparable three-stage rocket, the Saturn V, a great, fiery furnace of a beast. It was as big and heavy as a navy destroyer ship, and it could deliver 7.5 million pounds of thrust. What a long way these men had come since the days of the 300-, 1,000-, and 1,500-kilogram engines at Kummersdorf! This giant was 281 feet long and 33 feet in diameter. The 82-foot Apollo Command Module was designed to sit on top, in the same way the Explorer satellite had. From this position it would separate from the last of the three boosters, gaining sufficient speed to be thrust out of Earth orbit, streaking its way to the Moon.

The first Saturn V was launched November 9, 1967 from Launch Complex 39 at the Kennedy Space Center. The development was right on schedule, and the flight went without a hitch. Saturn V was ready to roll.

The first "live" Saturn V launch, with astronauts onboard, came almost a year later, when *Apollo 8* launched just before Christmas 1968. It was the third Saturn V to launch. In the Command Module were Commander Frank Borman, Command Module pilot James A. Lovell, Jr., and the Lunar Module pilot, William A. Anders.

This would be the first spacecraft to head out of Earth orbit toward the Moon. As the countdown completed, the big Saturn V booster produced its deafening roar and blinding flame. The big bird hovered for a moment and then slowly lifted. Within minutes the two spacecraft, still attached to the third-stage Saturn IV-B rocket, had achieved orbit around Earth. The astronauts orbited long enough to run a safety check on all systems, and then it was the job of the Saturn IV-B to power on and kick the Command Module-Lunar Module spacecraft into "translunar injection," sending it on its way to the Moon. This last stage of the rocket then dropped away, its duty completed and its fuel spent.

Almost miraculously, two days later the astronauts were broadcasting TV pictures and voices from an orbit around the Moon. They had made it!

In the long road leading up to the first Moon landing, two more test launches were scheduled. In February 1969 the *Apollo 9* crew went into Earth orbit to try out the docking and undocking of the Lunar Module close to home—so a rescue might be possible if anything went wrong. With so much combined experience, von Braun and the German rocket scientists knew how many things could go wrong in complex electronic and mechanical systems. Especially when human lives were aboard, they liked to play it safe. Redundancy—having a backup system to take over if there is a problem with the primary system—was a key principle they instituted in the space program, and time and again their caution proved wise.

The rest of the Apollo engineers learned this lesson the hard way, back on January 27, 1967, when the first Apollo crew climbed aboard the Command Module for a test-run simulation. Flawed wiring sparked the pure oxygen atmosphere in the module and engulfed the crew in flames almost instantly. Trapped inside, they all died.

The Command Module had suffered many problems with inferior parts and shoddy workmanship. Combined with insufficient safety precautions, these shortcomings had cost the lives of three men. The setback also was costly to the space program, and, like the explosion of the *Challenger* Shuttle spacecraft on January 28, 1986, the tragedy caused agonized rethinking on the part of everyone participating in the program.

The 1969 *Apollo 9* mission went smoothly, however, and on May 18 *Apollo 10* was rolled out on to the launch pad, ready for liftoff. This mission would test the Lunar Module in orbit around the Moon, without attempting a landing. The test went smoothly, leaving only one final challenge.

It was hot the morning of July 16, 1969, and the three astronauts—Neil Armstrong, Buzz Aldrin, and Michael Collins—lay on their backs inside the cramped Command Module as they waited for the countdown. Three miles away, in the launch control center, von Braun, launch operations director Kurt Debus, and the rest of the launch crew progressed point by point through the final countdown.

Lift-off. The blinding white exterior of the great Saturn V rocket shone like a beacon in the sunlight. Von Braun could see it out the window, poised on the launch pad, both mammoth and graceful, the apex of 45 years of work. The seconds ticked by. Everyone in the room felt the tension, the excitement, and the anxiety. At 9:32

July 16, 1969: *Apollo 11* officials show relief and joy at the successful countdown and liftoff. From left to right at the console: Charles W. Mathews, NASA deputy associate administrator for manned space flight; Wernher von Braun; George E. Mueller, NASA associate administrator for manned space flight; and Lieutenant General Samuel C. Phillips, director of the Apollo program. (NASA Marshall Space Flight Center Archives)

July 16, 1969: *Apollo 11* shoots for the Moon with three astronauts aboard.
(NASA Marshall Space Flight Center Archives)

A.M., a flame glowed beneath the huge engines, and slowly, slowly it rose up. The launch control center shook with the billowing shock wave from the rocket's roar. Now the rocket was lifting, pulling farther and faster away from Earth. Von Braun watched intently as it pulled upward, his eyes riveted to every contour, as if pushing the rocket upward with sheer willpower.

Then a noise jangled into his consciousness. The whole room had exploded with yells, laughter, and cries of joy—people shaking hands, clapping each other on the back. The mood of anxiety had suddenly burst into enormous relief and excitement. They had done it! Armstrong, Aldrin and Collins were on their way to the Moon, and the rocket team had provided the first big push.

On July 20 Armstrong and Aldrin climbed into the Lunar Module named Eagle, undocked it from the Command Module in which Collins remained as pilot, and backed it away. In one of the most exciting adventures of all time, they piloted the Lunar Module down to within 50,000 feet, then burned the descent rocket to brake their movement toward the Moon to come in for a soft landing. Finally the message came over the intercom at Mission Control in Houston, Texas. "Houston, Tranquility Base here," the voice crackled. "The Eagle has landed." Seven hours later Neil Armstrong stepped out on the surface of the Moon, the first human being ever to set foot on another celestial body. We had in fact flown out of our cocoon.

Five more successful landings on the Moon followed over the next three years. By the late 1980s Earth-orbit flights had become almost commonplace. And while interest in space waned after the early 1970s, space satellites, shuttles, and spacecraft have become a permanent part of our lives, interwoven with every aspect of our culture—a new culture that had begun in Germany years before, when a small boy sent for a book in the mail, a book whose equations and diagrams he trained himself to understand.

13

RETURN TO A VISION: NEXT STOP, MARS

Wernher von Braun left Huntsville, Alabama in February 1970 with mixed emotions. He was leaving the rocket team he had led for four decades. He was leaving a place that loved him, where people regarded him as a hero. In a lavish Huntsville farewell celebration, banners waved above the streets proclaiming: DR. WERNHER VON BRAUN—HUNTSVILLE'S FIRST CITIZEN—ON LOAN TO WASHINGTON.

Huntsville would never forget the bronzed, charismatic man with the sparkling eyes who had led the transformation of the sleepy little cow and cotton town into a booming center of rocket technology. Five years later the town dedicated a $14 million civic center named in von Braun's honor, with 2,000 people gathered for the ball thrown for the opening. There, Ernst Stuhlinger proposed a toast, his champagne glass lifted high, and spoke for the city that had become his home. He said:

> Twenty-five years ago, Wernher, when you and your co-workers lived in Texas, you told us one day that we would soon move to a place far in the East and deep in the South. You said the city of Huntsville had everything we could wish. There was plenty of room for our work, it was small, but a beautiful place to live, and all the people were extremely cordial.
>
> . . . we know that a man who is reaching for the stars cannot be held at one place on Earth for long, but we are proud that you made our city your home for twenty years. Those years certainly changed the face and the fate of the Earth . . .

But if, as his Huntsville friends proclaimed on that cold February day in 1970, von Braun initially left on loan from a place that loved him, he headed for a place that did not welcome him with open arms. His destination was Washington, D.C., where many people still regarded him with a mixture of envy and mistrust. And, of course, while he was, in essence, being promoted, he could expect to have far less influence and power as a deputy administrator at headquarters than he had enjoyed as director of one of NASA's centers. He would not, he knew, be groomed for the position of NASA administrator. While he had found a home and fame in America, he still carried with him the shadowy ghost of his past.

As Charles S. Sheldon, Ph.D., former White House staff member on the National Aeronautics and Space Council, would later remark: "There was always a lingering resentment at the Washington end toward von Braun and his team. There were always rumors that von Braun would someday be head of NASA. But there is a great sensitivity in Washington about racial and ethnic interests. . . . Von Braun would never be given a political position." Those who had not forgotten the Nazi extermination of millions of innocent people would not welcome a former member of the Nazi party into the seat of the U.S. Federal Government.

But finally NASA administrator Thomas O. Paine persuaded von Braun to accept the new post. Richard Nixon, then President of the United States, wanted to initiate a new space program—one to which his name could be attached, as Kennedy's had to the Apollo program, a new program of vision and scope. Even before Neil Armstrong and Buzz Aldrin had landed on the Moon, Nixon had formed a Space Task Group to project the next steps for NASA, and in September 1969 the group published a book called *The Post-Apollo Space Program: Directions for the Future.* About the same time NASA came out with *America's Next Decades in Space: A Report of the Space Task Group,* and a third publication, *The Next Decade in Space,* came out in March 1970 from the President's Science Advisory Committee. All three outlined an ambitious program aimed at a voyage to Mars by 1989. This was the kind of plan that von Braun could throw his energies into.

"Yes—we've walked on the Moon," he would say. "Now let's get the funds to go on to Mars."

They were ambitious and exciting plans, and they looked good on paper. But many, both inside and outside of NASA, had deep doubts that they would ever see fruition. While the more starry-eyed of the space enthusiasts felt their hearts quicken with enthusiasm and hope, others, by the early 1970s, were beginning to feel that somehow a wonderful dream was over, that the optimistic magic of the '60s had given away to the cold economic reality of the '70s. For many dedicated space program watchers, the glowing optimism of the Space Task Force and NASA reports was numbingly depressing, so "unreal" in the light of political, social, and economic reality, that it made the achievements of the '60s seem almost equally unreal. Did the President of the United States actually set a public goal of landing a man on the Moon? Did an enthusiastic public applaud with enthusiasm and follow with baited breath each step along the way? Were wonderful rockets built and sent, and wonders returned? Or had it all been some science fiction magazine story, a dream that vanished too quickly upon awakening?

The world had changed, the Kennedy years gone, the peak of the space era over, and it had all seemed so quick, so brief after so many years of waiting and dreaming. The "dreamers of the future" suddenly and ironically seemed like anachronistic relics of the past. These were not good days for von Braun. Usually able in the past to parlay his imagination, drive, practicality, and wit into viable, funded programs, now he seemed more and more like a useless spirit of another era. Where once he had been a symbol of what could be, now he seemed more a ghost of what once was—or worse, what almost was.

To make matters more difficult, on July 28, 1970 Thomas Paine resigned as NASA administrator to go to work for the General Electric Company.

Without Paine, who had spearheaded Nixon's Space Task Group, Nixon's flagging interest in space, drained by the growing political and economic problems facing his administration, waned even further. A new generation of NASA administrators greeted von Braun's efforts to breathe life into the mission to Mars with faint enthusiasm. By the end of 1969 Nixon had announced that he would like to see a man on Mars by 1982, but his heart probably was never really in it.

Von Braun visited India in 1975 to discuss educational uses of the ATS-6 communications satellite. Pictured here is one of the schools in India that benefited from the program. (Archives, U.S. Space and Rocket Center, Huntsville, Alabama)

In 1971—the year that von Braun had once enthusiastically predicted would see a permanent manned base on the Moon—as NASA prepared for difficult budget hearings in Congress, the former director of Marshall Space Flight Center, now 59, was often seen walking the halls alone. His charismatic boyishness and energy were missing, and he looked defeated and depressed as he emerged from planning sessions with the new administrator, James C. Fletcher.

On June 10, 1972 an era and a lifelong dream came to an end when Wernher von Braun retired from NASA to work for the first time in private industry—as vice president for engineering and development at Fairchild Industries, a major aerospace engineering company with headquarters in nearby Germantown, Maryland.

Fairchild had some forward-looking ideas, and the move fit well with von Braun's visionary style. Fairchild had built an Applications Technology Satellite, known as ATS-6, which the United States planned to lend to India for a year in 1975 for a massive educational project. TV transmissions from the satellite would be beamed to schools in some 5,000 Indian villages, making educational material available to youngsters who otherwise had no access to the world beyond their immediate communities. The government of India was in the middle of planning for an even bigger, long-range project of its own that could have a much more widespread effect—a system that could reach 500,000 such villages. It hoped to make five TV channels available, with materials offered in 25 languages, for its multicultural country. Fairchild, a leader in satellite technology and ground equipment, hoped to get the job.

The work required extensive traveling for von Braun, who, after years of speaking tours and NASA installation-hopping, thought nothing of making several trips in the space of days to locations as widespread as New Delhi, India, Tehran, Iran, and Washington, D.C.

He never ceased to stump for space, founding the National Space Institute (NSI; now the National Space Society) in 1974 to further the cause of space exploration.

In between consultations and speeches, he enjoyed his home in Alexandria, Virginia, where he and his family moved after leaving Huntsville. There he loved to swim in the pool and spend evenings in the luminous white observatory he had built in his backyard, spending time with the Celestron telescope friends had given him on his 60th birthday. Wernher von Braun continued to function nonstop.

In July 1975 von Braun traveled to the Kennedy Space Center to witness the launch of the last Saturn 1-B. Aboard, three astronauts—Frank Thomas Stafford, Vance Brand, and Deke Slayton—rode in an *Apollo* command capsule to meet and dock with a Soviet *Soyuz* spacecraft. In an unprecedented cooperative mission between the United States and the Soviet Union, Soviet cosmonauts Alexei Leonov and Valery Kubasov met in space with their American counterparts, orbiting together around the Earth in the historic *Apollo-Soyuz* "handshake in space."

Elated at the liftoff, von Braun confessed to CBS-TV news commentator Walter Cronkite, "I couldn't help but say 'Go, baby, go!' when ignition occurred." That same day von Braun rushed off to Stuttgart, Germany to accept election to the board of directors of Daimler (Mercedes) Benz.

Then, in the midst of von Braun's energetic work with Fairchild and the NSI, he learned that he had cancer. Surgery failed to stem the progress of the disease, and ill health forced him to retire on December 31, 1976. President Gerald R. Ford awarded him the

National Medal of Science just before he left office in early 1977, and it was presented to him in the hospital by Edward G. Uhl, chairman of Fairchild Industries. Tears came to the eyes of the man who, by his own description, came to this country with all his belongings in one pasteboard carton. It was a fitting final tribute from a nation who had made use of the considerable talents of a man who had become, in one lifetime, both a despised perpetrator of war and arguably the greatest proponent of space exploration.

As Harry Stine wrote in his book *ICBM:*

> Personally, Wernher von Braun was not only a consummate diplomat but also one of the best engineers I've ever known. He was a role model. A man with dreams *and* academic credentials, a person who could theorize *and* design, an individual whose charisma was so powerful that he could instantly dominate any assemblage of people by his mere presence *and* who was also warm and friendly, he was also a practical person who knew how to work with his hands.

Wernher von Braun died on June 16, 1977 in Alexandria, Virginia. He was 65 years old.

It was a day, unquestionably, for remembering his considerable contributions to the expansion of human knowledge and experience. In the words of U.S. President Jimmy Carter, "To millions of Americans, Wernher von Braun's name was inextricably linked to our exploration of space and to the creative application of technology. . . . Not just the people of our nation, but all the people of the world have profited from his work."

CHAPTER 13 NOTES

p. 109 "Twenty-five years ago, Wernher . . ." Erik Bergaust, *Wernher von Braun* (Washington, D.C.: National Space Institute, 1976), p. 375.

p. 110 "There was always a lingering . . ." Quoted by Frederich I. Ordway, III, and Mitchell Sharpe in *The Rocket Team* (New York: Thomas Y. Crowell, 1979), pp. 453–454.

p. 110 "Yes—we've walked on the Moon . . ." Bergaust, p. 548.

p. 113 "I couldn't help . . ." Bergaust, p. 526.

p. 114 "Personally, Wernher von Braun . . ." Harry G. Stine, *ICBM: The Making of the Weapon that Changed the World* (New York: Orion Books, 1991), pp. 23–24.

p. 114 "To millions . . ." Frederick I. Ordway and Mitchell R. Sharpe, *The Rocket Team* (New York: Thomas Crowell Publishers, 1979), p. 456.

14

EPILOGUE: THE LEGACY OF WERNHER VON BRAUN

Wernher von Braun never failed to be elated whenever he saw one of his huge "birds" soar upward from the Earth's crust toward what his colleagues came to call "von Braun's beloved stars." To him a rocket was a work of wonder, not one man's work but the result of consummate, tireless teamwork and a dedication to excellence, characterized, as he once wrote, "by enthusiasm, professionalism, skill, imagination, a sense of perfectionism, and dedication to rocketry and space exploration."

But thundering rockets are only a portion of the legacy Wernher von Braun left. The United States depended heavily on von Braun's know-how, vision, and leadership during the intense Apollo Project years, and even after his days at NASA, the space rockets that he had made possible continued to affect our lives—both in their military potential and in their scientific and commercial applications.

He had opened up the "space age" and, in so doing, precipitated, for a while at least, one of the most exciting periods in the history of human knowledge. Driven by national pride and military expediency, both the United States and the Soviet Union pushed heavily toward the exploration and use of space in the 1950s, '60s, and '70s. Every move that the United States made led to a move by the Soviet Union, and vice versa, each country always striving to get ahead or stay ahead in the "space race."

Between 1958 and 1976 the Soviet Union and the United States sent 80 missions—mostly unmanned—to the Moon. Both countries obtained vast amounts of new information, including the first photographs of the Moon's far side (which is always turned away from Earth). Probes to Venus, Mars, Mercury, and the far-flung outer planets soon followed both from Cape Canaveral (called Cape Kennedy 1963–1973) and the Soviets' Baikonur launch site in Kazakhstan. The U.S. Pioneer and Voyager series were among the most stellar. As these roaming robots snapped images and sent signals earthward, they collected data that revolutionized human understanding of the Solar System and the Universe beyond.

By 1971 the Soviet Union had begun a marathon of experiments on living in space. Soviet "cosmonauts" spent ever-longer stays—up to a year or more—in a series of space stations culminating in the big Mir ("peace") station, launched in 1986. The United States, meanwhile, sent three teams of astronauts in 1973–74 to a space station called Skylab to study the Sun. Skylab itself was launched by a Saturn V rocket, and Apollo technology was used to launch the astronauts as well.

A Saturn rocket was also used to launch three United States astronauts into orbit aboard an Apollo Command Module, which on July 18, 1957 completed the historic docking maneuver with a Soyuz spacecraft piloted by two Soviet cosmonauts.

Then, in 1982, the United States launched the first of a fleet of Earth-orbiting vehicles called Space Shuttles or, officially, the Space Transportation System. Already a victim of waning congressional interest in the 1970s, however, the shuttle design was patched together after numerous budget cutbacks and was a far cry from the advanced designs envisioned by von Braun. Badly compromised in its design and mission, the U.S. Space Shuttle finally flew, making its first operational flight in November 1982. And, with the exception of a 2½-year hiatus following the tragic explosion in 1986 of the *Challenger,* the shuttle has proved a trusty vehicle for short-term scientific experiments in space as well as complicated satellite launchings and repairs, often done by astronauts floating in the great expanse of the Universe with nothing more than a spacesuit between them and the vast vacuum of space.

In addition to deployment of countless satellites having a variety of purposes ranging from the scientific to the commercial, the space programs of both the United States and the Soviet Union have collected extensive data about the effects of weightlessness on living organisms (including humans), crystal formation, and countless other areas of inquiry. By the 1980s many other countries, including China, Japan, India, and a consortium of European nations, had obtained the capability to launch satellites and had developed space programs of their own.

Today's vast communications web—from complex television networks and "superstations" to the "Information Superhighway," from sophisticated weather reporting to "instant" news and communications transfers—couldn't exist without the satellite links that have truly made the world a global community.

All this was, in a way, positive fallout from the Cold War between the United States and the Soviet Union. What to Tsiolkovsky, Goddard, Oberth, and von Braun seemed inevitable might never have happened when it did if two countries hadn't been vying to one-up each other in the eyes of the world. But as the Cold War drew toward an end in 1989, and with the breakup of the Soviet Union in 1991, some of the spur for this surge of interest in the use and exploration of space dissipated. With both the United States and Russia facing serious financial and social problems, both nations began to cut back on their space programs. Among the first casualties, the Mir space station limped along, neglected, in the early 1990s, and the U.S. Congress reduced funds for a planned international space station called Freedom.

Visionaries in both nations, though, still plan and dream. Plans for a joint manned mission to Mars using both Russian and American technology have been brewing for years—and von Braun, one of the first to envision such a voyage, would have been elated at the prospect. With friendlier relations between the nations, the project is possible, but without a public mandate, funding may be scarce, with both countries more likely to funnel their resources toward urgent needs on Earth.

While the hard realities of the world may have dimmed some dreams of space travel and exploration, the fact remains that, thanks to the visions and work of Wernher von Braun and others, humankind today is not only able to explore space, but grows increasingly dependent on its continued commercial exploitation. Rockets and satellites have found a permanent place in our culture;

In a 1969 photo, Wernher von Braun stands beside the fulfillment of his dream to build a rocket that could take people to the Moon: the giant Saturn V rocket. (NASA)

the future as foreseen by Tsiolkovsky, Goddard, Oberth, and von Braun is here.

Von Braun created, through the campaign he waged and won for space, a new vision, a wider view, not just of the Earth but of the

Universe. He taught that we could crawl out of the Devonian sea of Earth's atmosphere, to zoom into the vast, infinite reaches of space. He believed that humankind could emerge from its cocoon and fly, like his rockets—using his rockets—into unknown realms of knowledge and freedom.

But he also left another, more troubling legacy, one of a man whose vision of possibilities and opportunities at times obscured other, more human values, values of integrity and conscience.

Twenty-five years after the end of World War II, von Braun wrote:

> What is a man's duty who loves his native country, sees that it has fallen into evil hands, but also that its very survival is at stake? It is easy to answer this question from the safe shelter of a strong, well-governed nation, which is more or less at peace with the rest of the world. The answer does not come as readily when you live right in the midst of such a storm of events.

But he should, perhaps, have asked two other questions as well: How do we balance our obligations as human beings against our duties to our government or country? and When do we have an obligation to say "No" and act against crimes we see committed around us?

Did von Braun fall short during his years at Peenemünde, in Nazi Germany? Could he have acted differently, or should he have? In his own words:

> I have come to the conclusion that it is better for a man to respect another man's conscience than to sit in glib judgment of him. The final judgment on matters like this is surely in better hands with the good Lord. I thought, and I still do, that under the circumstances, I did what was my duty.

Whether, in the end, this question of conscience outshines all other questions about Wernher von Braun is open to debate. But his life raises the issue, and thereby leaves a second, darker legacy. During his last 30 years, he captured the Moon, the stars, Jupiter, and a skyful of planets—a stellar gift. His earlier years also provoke thought, forcing a consideration of the complexity of human ethical decisions and their consequences.

Earthrise as seen from the Moon by the *Apollo 11* crew, July 1969. (NASA)

This two-pronged heritage—offering wide-open possibilities and posing deep-seated questions—epitomizes Wernher von Braun's mark upon our time.

CHAPTER 14 NOTES

p. 116 "von Braun's beloved stars." George Barrett, *The New York Times Magazine,* October 20, 1957, p. 87.

p. 116 "by enthusiasm . . ." Von Braun, in his foreword to Frederick I. Ordway and Mitchell R. Sharpe, *The Rocket*

Team (New York: Thomas Crowell Publishers, 1979), p. xii.

p. 120 "What is a man's duty . . ." From a letter to R. W. Reid, quoted in Reid, *Tongues of Conscience: Weapons Research and the Scientists' Dilemma* (New York: Walker and Co., 1969) p. 116.

p. 120 "I have come to the conclusion . . ." Quoted by Reid, p. 116

GLOSSARY

A-4: The most advanced rocket developed to completion by the rocket team at Peenemünde; also known as the V-2.

ballistic missile: A projectile that falls in a free-falling flight pattern after a guided, self-powered ascent.

ballistic trajectory: The curved path of a projectile, such as a missile or a bullet in free fall, after the thrust or propelling force has ended.

booster: In a multistage rocket, the rocket that provides an early stage of propulsion, including the initial stage that provides power for the launching and first part of the flight.

Cold War: After World War II, an extended period of animosity between the Soviet Union and the United States, when the countries were hostile to and uncooperative with each other, even though not formally at war.

combustion: Rapid oxidation or burning.

gyroscope: A device usually consisting of a spinning mass (e.g., a disk or wheel), suspended so that it maintains the same orientation no matter which way it is turned, as long as it is not subjected to external changes in velocity. A gyroscope can be used to automatically keep a craft in the proper relative position or at the proper attitude.

launch vehicle: Another term for rocket.

Luftwaffe: The German air force during the Nazi regime.

missile: Usually a rocket carrying a warhead, or bomb, as a payload.

ordnance: All supplies pertaining to weapons and their maintenance, including the weapons themselves.

payload: Anything transported by a rocket or spacecraft that is not involved in its primary functions.

projectile: Any object propelled forward by a force, for example, a bullet or a rocket.
propulsion: A force that propels, or pushes, an object forward.
rocket: A device propelled by ejection of matter, especially by the ejection of gases resulting from combustion.
rocket engine: An engine that produces propulsion, by ejection of matter, especially by the ejection of gases resulting from combustion; the motor portion of a rocket.
sounding rocket: A rocket used to obtain information about the atmosphere.
static test: (Or static firing.) A test in which the rocket motor is fired, but kept stationary and not allowed to leave the test stand, providing an opportunity to measure engine thrust and other performance characteristics.
upper stage: In a multistage rocket, a booster rocket that takes over after the first-stage booster has burned its fuel.
V-2: Hitler's "Vengeance Weapon #2," originally known as the A-4 rocket.
Wehrmacht: The German army, especially during the Nazi regime.

FURTHER READING

Books Written by Wernher von Braun:

Ley, Willy, and Wernher von Braun. *The Exploration of Mars.* New York: Viking Press, 1956.

von Braun, Wernher. *First Men to the Moon.* New York: Holt, Rinehart and Winston, 1960.

———. *The Mars Project* (translation of *Das Marsprojekt,* Frankfurt am Main, 1952). Urbana: University of Illinois Press, 1953, 1962, and 1991.

von Braun, Wernher, Silvio A. Bedina, and Fred L. Whipple. *Moon: Man's Greatest Adventure.* New York: Harry N. Abrams, 1973.

von Braun, Wernher, and Frederick I. Ordway III. *The Rockets' Red Glare: An Illustrated History of Rocketery Through the Ages.* New York: Doubleday/Anchor Press, 1976.

von Braun, Wernher, Frederick I. Ordway III, and Dave Dooling. *Space Travel: A History.* New York: Harper & Row, 1985.

von Braun, Wernher, Fred Whipple, and Willy Ley. *Conquest of the Moon,* ed. Cornelius Ryan. New York: Viking Press, 1953.

Books and Articles About Wernher von Braun:

Barrett, George. "Visit with a Prophet of the Space Age." *The New York Times Magazine,* October 20, 1957, pp. 14ff.

Bergaust, Erik. *Reaching for the Stars.* New York: Doubleday, 1960.

———. *Wernher von Braun.* Washington, D.C.: National Space Institute, 1976.

David, Heather M. *Wernher von Braun.* New York: G. P. Putnam's Sons, 1967.

Goodrum, John C. *Wernher von Braun.* Tomball, Tex.: Strode Publishers, 1969.

Lampton, Christopher. *Wernher von Braun.* New York: Franklin Watts, 1988.

Stuhlinger, Ernst, with Frederick I. Ordway III. *Wernher von Braun, Crusader for Space: A Biographical Memoir.* Malabar, Fla.: Krieger Publishing Co., 1994.

Walters, Helen B., *Wernher Von Braun: Rocket Engineer.* New York: Macmillan Company, 1964.

Books and Articles About Rocket Development and Space:

Of the many fine books on rockets and space exploration, we list just a few of the classics and a few of the most recent and insightful. In addition to *We Seven,* many astronauts have written their own accounts of their space experiences.

Baker, David. *The Rocket.* New York: Crown Publishers, 1978.

Burrows, William E. *Exploring Space: Voyages in the Solar System and Beyond.* New York: Random House, 1990.

Carpenter, M. Scott, et al. *We Seven.* New York: Simon & Schuster, 1962. The story of the first *Mercury Seven* astronauts.

Clarke, Arthur C. *Astounding Days: A Science Fictional Autobiography.* New York: Bantam Books, 1990.

Dornberger, Walter. *V-2.* New York: Viking Press, 1954.

Garlinski, Jozef, *Hitler's Last Weapons.* New York: Times Books, 1978.

Gatland, Kenneth. *Fact Finder Space Diary.* New York: Crescent Books, 1989.

Goddard, Robert H. *A Method of Reaching Extreme Altitudes.* Washington, D. C.: Smithsonian Institution, 1919.

Gray, Mike. *Angle of Attack: Harrison Storms and the Race to the Moon.* New York: W.W. Norton and Company, 1992.

Holder, William G. *Saturn V: The Moon Rocket.* New York: Julian Messner, 1969.

Hunt, Linda. "U.S. Coverup of Nazi Scientists." *The Bulletin of the Atomic Scientists,* April 1985, Vol. 41, No. 4, pp. 16–24.

Huzel, Dieter K., *Peenemünde to Canaveral.* Englewood Cliffs, N.J.: Prentice-Hall, 1962.

Lehman, Milton. *This High Man: The Life of Robert H. Goddard,* with a preface by Charles A. Lindbergh. New York: Farrar, Straus and Co., 1963.

Ley, Willy. *Rockets, Missiles, and Men in Space.* New York: Viking Press, 1968.

———. *Satellites. Rockets and Outer Space.* New York: Signet Key Books, 1958.

Manno, Jack. *Arming the Heavens.* New York: Dodd, Mead and Company, 1984.

McDougall, Walter A. *. . . the Heavens and the Earth.* New York: Basic Books, 1985.

McGovern, James. *Crossbow and Overcast.* New York: William Morrow, 1964.

Murray, Charles, and Catherine Bly Cox. *Apollo: The Race to the Moon.* New York: Simon and Schuster, 1989.

Ordway, Frederick I., and Mitchell R. Sharpe. *The Rocket Team.* New York: Thomas Crowell Publishers, 1979.

Reid, R. W. *Tongues of Conscience: Weapons Research and the Scientists' Dilemma.* New York: Walker and Company, 1969.

"The Seer of Space," *Life* magazine, November 18, 1957, pp. 133–139, including an interview with Wernher von Braun ("Plain Talk from von Braun"), by Richard B. Stolley, p. 136.

Stine, Harry G. *ICBM: The Making of the Weapon That Changed the World.* New York: Orion Books, 1991.

Taylor, Theodore. *Rocket Island.* New York: Avon Books, 1985. Brief, interesting story of Peenemünde told from the military rather than the technical standpoint. Not much new but easy to read.

Tsiolkovsky, Konstantin E., *Beyond Planet Earth,* translated from the Russian by Kenneth Syers. Originally published in Russian as *Exploration of Space by Rocket Devices* in 1903. New York: Pergamon Press, 1960.

———. *Selected Works,* translated from the Russian by G. Yankovsky. Moscow: Mir Publishers, 1968.

Winter, Frank H. *Prelude to the Space Age: The Rocket Societies: 1924–1940.* Washington, D.C.: Smithsonian Institution Press, 1983.

Wolfe, Tom. *The Right Stuff.* New York: Farrar, Straus and Giroux, 1979. Bantam paperback edition, 1980.

Books About Germany During Hitler's Regime:

Speer, Albert. *Inside The Third Reich.* New York: Macmillan Co., 1970. Among hundreds of excellent books, this one gives a telling inside view.

INDEX

Illustrations are indicated by *italic* numbers.
The letter *g* after a number indicates a word in the glossary.